Springer Praxis Books

This book series presents the whole spectrum of Earth Sciences, Astronautics and Space Exploration. Practitioners will find exact science and complex engineering solutions explained scientifically correct but easy to understand. Various subseries help to differentiate between the scientific areas of Springer Praxis books and to make selected professional information accessible for you.

More information about this series at http://www.springer.com/series/4097

Dmitry M. Ermakov

Satellite Radiothermovision of Atmospheric Processes

Method and Applications

Springer

Published in association with
Praxis Publishing
Chichester, UK

Dmitry M. Ermakov
Fryazino Branch of the Kotelnikov Institute
of Radioengineering and Electronics
Russian Academy of Sciences
Fryazino, Russia

Space Research Institute
Russian Academy of Sciences
Moscow, Russia

Springer Praxis Books
ISBN 978-3-030-57084-2   ISBN 978-3-030-57085-9 (eBook)
https://doi.org/10.1007/978-3-030-57085-9

This Springer imprint is published by the registered company Springer Nature Switzerland AG
The registered company address is: Gewerbestrasse 11, 6330 Cham, Switzerland

Preface and Acknowledgements

This book is based on materials from a series of works initiated by my acquaintance and beginning of cooperation with Prof. E. A. Sharkov from the Space Research Institute of the Russian Academy of Sciences. It was he who put forward the idea of a systematic search for the relationship between the evolution of tropical cyclones and the large-scale dynamics of the total precipitable water (TPW) field. The first stage of the work was devoted to the design and implementation of a new database of satellite radiometric observations and the development of tools for their visualization and preliminary analysis (2007). It became clear that the traditional mosaic representation of satellite swaths with gaps in the tropical zone and a twelve-hour time step does not satisfy the research objectives.

The second crucial step in the development was the discussion of the problem with Dr. A. P. Chernushich from the Institute of Radioengineering and Electronics of the Russian Academy of Sciences. He proposed the use of a motion estimation and compensation method for spatiotemporal interpolation of source satellite data (2010). Eventually, a compact team of the three researchers was formed, the efforts of which produced almost all of the results summarized in the book.

From this moment on, work has begun at an increasing pace. In 2011, by trial and error, the optimal algorithm for constructing reference fields was established—the initial step in dynamic analysis. In 2012, we presented our first detailed dynamic pictures of the evolution of several tropical cyclones (including Katrina) in the TPW field. By the end of 2013, the concept of quantitative analysis was formulated and brought to software implementation, based on the calculation of the power of latent heat fluxes. The following year, the technique was successfully applied to a set of all tropical cyclones that reached the maximum category in August 2000. At about this point, we replaced the original name of the approach ("animation analysis") with "satellite radiothermovision." In 2015, the technique was generalized to the case of the analysis of large-scale synoptic processes, and in 2016 the geoportal of satellite radiothermovision was launched. In 2017, on the basis of a generalized technique, the first results on the study of global atmospheric circulation parameters were obtained, deepened in subsequent years. At the same time, the first studies of the so-called atmospheric rivers were started.

The creation of this book would not have been possible without the participation of a large number of people. A key contribution to the development of the satellite radiothermovision method is made by my colleagues and permanent co-authors of scientific publications, Prof. E. A. Sharkov and Dr. A. P. Chernushich. Professor B. G. Kutuza and Dr. V. P. Savorskiy actively supported and expressed constant interest to the progress of the work. Professor K. P. Gaykovich, Prof. I. A. Repina, Dr. E. N. Kadygrov, Dr. A. F. Nerushev, and Dr. M. T. Smirnov got acquainted with the significant parts of the book at various stages of its preparation and made a number of valuable comments. Dr. L. M. Mitnik attracted the author's attention to some independent conceptually close researches in the field, mentioned in this book, and to the problem of studying of atmospheric rivers.

The author is grateful to Clive Horwood (Praxis Publishing) for the enthusiasm and warm personal participation in the organization of the project, as well as to Christian Witschel's team from Springer for their assistance in the process of its implementation.

Sincere thanks to my family for the patience and encouragement that I felt all the time working on the book, especially during the "quarantine era" caused by the COVID-19 pandemic.

In conclusion, the author would like to mention several organizations that provide free access to satellite data and processing products used in his research.

Regarding the fields of geophysical parameters of the atmosphere and ocean reduced to a regular geographic grid of 0.25 degrees, SSM/I, SSMIS, AMSR-E, AMSR-2, WindSat, MW OI SST data are produced by Remote Sensing Systems and sponsored by NASA. Data are available at www.remss.com.

Regarding the TPW fields on the original grid of AMSR-2 observations, GCOM-W (AMSR2) data by Japan Aerospace Exploration Agency were used.

References to data sources are given in the relevant sections of the book.

Fryazino/Moscow, Russia Dmitry M. Ermakov

About This Book

Satellite Radiothermovision of Atmospheric Processes: Method and Applications is a book demonstrating the power of synergetic interdisciplinary research of the Earth's complex environmental systems. It combines the passive microwave remote sensing paradigm with fundamental knowledge of mesoscale and synoptic atmospheric processes and state-of-the-art approaches of big data processing in computer science including various machine vision techniques.

Continuous passive microwave (radiothermal) satellite monitoring of the Earth gives a strikingly high amount of detailed information on the dynamics of climatic subsystems, within which the Earth's atmosphere demonstrates probably the most noticeable responses to climate variations on a set of time scales from days to decades. One particular kind of atmospheric responses is change of frequency and intensity of potentially hazardous events like tropical cyclones and atmospheric rivers. A less prominent but planetary important one in a long run is the change in characteristics of the global atmospheric circulation.

Investigation of all these phenomena is heavily based on numerical modeling. However, it is well known that such modeling gives a variety of possible scenarios very sensitive to small disturbances in initial conditions. Hence, the advance in quality of critically important forecasts of the environmental dynamics is toughly linked to the progress in interpretation of observational data.

The book provides a detailed introduction to the original approach of satellite radiothermovision which allows retrieving atmospheric dynamics and latent heat advection based on a calculation scheme closed in respect of satellite radiometry data. When applied to a particular atmospheric process, e.g., tropical cyclone, the approach allows for retrieving its dynamic characteristics, estimating the convergence/divergence of latent heat, and relating latent heat fluxes with changes in process intensity (like rapid intensification) during its evolution. At planetary scales, the approach provides an effective means to calculate statistical, spatiotemporal, dynamic, and energy characteristics of synoptic processes, e.g., atmospheric rivers. Successful retrieving of many fundamental parameters of global atmospheric circulation convincingly proves the consistency of the proposed new approach when applied to big data arrays (about 17 years of continuous satellite observations).

In consonance with its main purpose, the book contains a problem-oriented description of specific mesoscale and synoptic atmospheric processes, a brief history of remote passive microwave investigations of the atmosphere, and an overview of origins and current state of data processing techniques aimed at their dynamical analysis. Particular chapters are devoted to practical applications concerning investigation of tropical cyclones, atmospheric rivers, and global circulation. A special chapter provides detailed information on the geoportal of satellite radiothermovision developed under the author's guidance and designed to provide the international scientific community with the products and procedures of dynamic analysis of operational and archived remote sensing data.

Contents

1 Introduction . . . 1
- 1.1 Weather, Climate and the Water Vapor . . . 1
- 1.2 Satellite Microwave Radiometry of the Atmosphere . . . 3
  - 1.2.1 *What* Do We See? . . . 3
  - 1.2.2 *Where* Do We See? . . . 5
  - 1.2.3 *When* Do We See? . . . 7
  - 1.2.4 What *Don't* We See? . . . 8
  - 1.2.5 A Broader View in Conclusion . . . 9

2 Why Satellite Radiothermovision? . . . 13
- 2.1 Elements of Mesoscale and Synoptic Atmospheric Dynamics . . . 14
  - 2.1.1 Mesoscale and Synoptic Atmospheric Processes . . . 14
  - 2.1.2 Mesoscale: Tropical Cyclones . . . 15
  - 2.1.3 Synoptic Scale: Atmospheric Rivers . . . 17
  - 2.1.4 Planetary Scale: Global Atmospheric Circulation . . . 19
  - 2.1.5 The Commonality of Research Objects of Various Scales in the Context of Satellite Radiothermovision . . . 20
- 2.2 Passive Microwave Remote Sensing of the Atmosphere in Short . . . 21
  - 2.2.1 Brief Historical Overview . . . 23
  - 2.2.2 Present State . . . 28
- 2.3 Instruments and Data to Our Service . . . 31
- 2.4 The Idea of "Dynamic" Data Analysis . . . 33
- 2.5 Concluding Remarks to this Chapter . . . 34
- References . . . 35

3 Fundamentals of Satellite Radiothermovision . . . 41
- 3.1 Physical Foundations . . . 42
- 3.2 Mathematical Apparatus . . . 45
  - 3.2.1 Optical Flow Analysis Problem . . . 45

3.2.2 Optical Flow Analysis in Earth Remote Sensing Problems . . . 50
3.2.3 The Bases of the Satellite Radiothermovision Approach . . . 51
3.3 Algorithms and Software Realization . . . 52
3.3.1 Constructing Reference Fields: Transformation to a Regular Grid . . . 55
3.3.2 Constructing Reference Fields: Lacunae Stapling . . . 58
3.3.3 Spatiotemporal Interpolation . . . 63
3.3.4 Some Brief Notes on Software Implementation . . . 66
3.4 Accuracy Analysis . . . 67
3.5 Iterative Extension of the Basic Scheme: A Multisensory Approach . . . 76
3.6 New Opportunities in the Remote Sensing . . . 79
3.6.1 Joint Analysis of Independent Satellite Measurements . . . 80
3.6.2 Ensuring Spatial Connectivity of Fragmented Observations . . . 80
3.6.3 The Study of Vector Fields of Advection . . . 81
3.6.4 Calculation of the Integral Characteristics of Mass and Energy Transfer . . . 81
3.7 Concluding Remarks to this Chapter . . . 84
References . . . 84

4 Satellite Radiothermovision of Tropical Cyclones . . . 89
4.1 Energy Characteristics and Energy Balance of a Tropical Cyclone . . . 90
4.1.1 Tropical Cyclone Energy Characteristics . . . 90
4.1.2 Tropical Cyclone Energy Balance Factors . . . 91
4.2 General Characteristics of the Data Used and Analysis Methods . . . 93
4.3 Evolution of Tropical Cyclones in the Field of Total Precipitable Water . . . 95
4.4 Complex Analysis in the Fields of Several Geophysical Parameters . . . 105
4.5 Expanding the Approach for Exploring a System of Interacting Typhoons . . . 109
4.6 Concluding Remarks to this Chapter . . . 117
References . . . 117

5 Satellite Radiothermovision of Atmospheric Rivers . . . 121
5.1 Problems of Detecting Atmospheric Rivers . . . 122
5.1.1 Data Gaps . . . 123
5.1.2 Setting Detection Criteria . . . 124
5.1.3 Advection Field Accounting . . . 125
5.1.4 Satellite Data Synchronization . . . 126

5.2 Synthesis of an Algorithm for Automatic Detection of Atmospheric Rivers 127
5.2.1 Specification of the Boundaries of the Basin 128
5.2.2 Discrimination of Air Masses of Middle Latitudes 130
5.2.3 Morphological Analysis 134
5.2.4 Combining the Fragments 140
5.2.5 Pruning Branches 140
5.3 Retrieval of Atmospheric River Characteristics 142
5.3.1 Analysis of Latent Heat Fluxes 142
5.3.2 Analysis of AR Images in Dynamics 144
5.3.3 Joint Analysis in the Fields of Several Geophysical Parameters of the Atmosphere 146
5.3.4 Joint Analysis Over the Ocean and Land 146
5.4 Concluding Remarks to this Chapter 149
References 149

6 Satellite Radiothermovision of Global Atmospheric Circulation 151
6.1 Characteristics of the Problem in the Light of Satellite Radiothermovision 152
6.2 Used Data and Analysis Technique 154
6.3 Analysis of the Retrieved Global Circulation Characteristics 156
6.4 Concluding Remarks to this Chapter 169
References 172

7 Welcome to the Geoportal of Satellite Radiothermovision 175
7.1 The Concept of Geoportal of Satellite Radiothermovision and the Network Service ICAR 175
7.2 Geoportal of Satellite Radiothermovision: Description of Data and Services 177
7.2.1 External Data Sources 179
7.2.2 Reference Collection 180
7.2.3 User-Level Products 181
7.2.4 User-Level Product File Output Format 181
7.2.5 Network Services 182
7.3 ICAR Network Service: Remote Processing of Virtually Integrated Data 183
7.3.1 Interface Solution 183
7.3.2 Syntax and Semantics of ICAR 186
7.3.3 Software Implementation 189
7.3.4 Data Provisioning and Functional Content 191
7.4 The Practice of Using Implemented Software Solutions 191
7.5 Concluding Remarks to this Chapter 192
References 193

8 Conclusion 195
8.1 Improving the Quality of Dynamic Analysis Products 196
8.2 Enhanced Spatial Coverage and Improved Detail 197
8.3 Extension of Analysis Time Interval 197
8.4 Extension of the Dimensionality of the Task 198

About the Author

Dmitry M. Ermakov was born on August 12, 1976. He graduated from Moscow Institute of Physics and Technology (MIPT) as a MS in applied mathematics and physics in 1998 and passed his postgraduate practice in the Kotelnikov's Institute of Radio-engineering and Electronics of Russian Academy of Sciences (IRE RAS) in 1998–2001 in the field of remote sensing. In 2002, he obtained Ph.D. degree in radiophysics from IRE RAS. His Ph.D. thesis was devoted to combined analysis of satellite remote optical and passive microwave data for study of ocean surface parameters.

Since that time, he is with Fryazino Branch of IRE RAS at the positions of a junior researcher, a researcher scientist, a senior researcher, and (currently) a leading researcher. For many years, he cooperates with the Space Research Institute (IKI RAS) and currently holds a part-time position of a leading researcher there. For several years, he also held a part-time position of senior researcher at the Research Center for Earth Operating Monitoring (NTs OMZ) of the Russian Space Systems. Since 2018, he is Associate Professor at the Department of Radio-electronic Systems of Location, Navigation and Communication of the Fryazino Branch of the MIREA—Russian Technological University.

In 2019, he obtained the Dr.Sc. degree (Russian state qualification) in radiophysics from IRE RAS. He has published over 100 scientific works. Among his main co-authors are Prof. E. A. Sharkov (IKI RAS),

Prof. B. G. Kutuza, Dr. A. P. Chernushich, Dr. V. M. Polyakov, Dr. V. P. Savosrkiy, and Dr. M. T. Smirnov (all—IRE RAS). He is awarded a Keldysh Medal of the Russian Federation of Cosmonautics.

Abbreviations

ACE	Accumulated Cyclone Energy
AMSR	Advanced Microwave Scanning Radiometer
AMSR-E	Advanced Microwave Scanning Radiometer for EOS
AMSU	Advanced Microwave Sounding Unit
AMV	Atmospheric Motion Vector
AO	Atlantic Ocean
AR	Atmospheric River
ATMS	Advanced Technology Microwave Sounder
CEOS	Committee on Earth Observation Satellites
CIMSS	Cooperative Institute for Meteorological Satellite Studies
DMSP	Defense Meteorological Satellite Program
ESMR	Electrically Scanning Microwave Radiometer
GCOM-W	Global Change Observation Mission for Water
GMI	GPM (Global Precipitation Measurement) Microwave Imager
HC	Hadley Cell
HSB	Humidity Sounder for Brazil
ICAR	Interactive Calculator for Atmospheric Research
ICZ	Intertropical Convergence Zone
IO	Indian Ocean
JAXA	Japanese Aerospace Exploration Agency
JMA	Japan Meteorological Agency
MHS	Microwave Humidity Sounder
MIMIC	Morphed Integrated Microwave Imagery at CIMSS
MJO	Madden–Julian Oscillation
MSMR	Multi-frequency Scanning Microwave Radiometer
MTVZA	Modul' Temperaturno-Vlazhnostnogo Zondirovaniya Atmosfery (Module of Temperature–Humidity Sounding of Atmosphere)
MW	Megawatt (1 MW = 109 Watt) or Microwave
MWHS	Microwave Humidity Sounder
MWRI	Microwave Radiation Imager

NASA	National Aeronautics and Space Administration
NEMS	Nimbus-E Microwave Sounder
NOAA	National Oceanic and Atmospheric Administration
NWP	Numerical Weather Prediction
OI	Optimally Interpolated
OSCAR	Observing Systems Capacity Analysis and Review Tool
PO	Pacific Ocean
PW	Petawatt (1 PW = 109 Megawatt = 1015 Watt)
RSS	Remote Sensing Systems
SAD	Sum of Absolute Differences
SCAMS	Scanning Microwave Spectrometer
SMMR	Scanning Multichannel Microwave Radiometer
SNPP	Suomi National Polar-orbiting Partnership
SSD	Sum of Squared Differences
SSEC	Space Science and Engineering Center
SSM/I	Special Sensor Microwave—Imager
SSM/T	Special Sensor Microwave—Temperature
SSMIS	Special Sensor Microwave—Imager/Sounder
SST	Sea Surface Temperature
STS	Strong Tropical Storm
TC	Tropical Cyclone
TD	Tropical Depression
TL	Tropical Low
TPW	Total Precipitable Water
TS	Tropical Storm
UTC	Temps Universel Coordonné/Coordinated Universal Time
WMO	World Meteorological Organization
WO	World Ocean

List of Figures

Fig. 1.1 Distribution of the total precipitable water in the Earth's atmosphere on September 2, 2019, the color scale of the values below; gray color—land; purple—data gaps (see explanations in the text) 4

Fig. 2.1 Classifications of atmospheric processes by their spatial scales 15

Fig. 2.2 Location of the measuring channels of some satellite microwave radiometers (arrows) relative to the spectrum of the optical thickness of the atmosphere in the range 1–200 GHz: empty arrows—measurements at nadir; with vertical/horizontal/circular shading—measurements on vertical/horizontal/circular polarizations during transverse scanning; shaded arrows—bipolarization measurements during transverse scanning. At the bottom of the graph is the spectral variation of the characteristic optical thickness of the standard atmosphere τ: the solid line is for tropical latitudes; long dotted line—for temperate latitudes in summer; short dash—for temperate latitudes in winter 25

Fig. 3.1 Dimension of objects and their real and perceived movements: **a** the position of the quasi-one-dimensional object (contrast boundary) at time t; **b** the position of the border at time $t + dt$, colored arrows show some possible options for translational movement; **c** rotation of a three-dimensional object around a horizontal axis, red arrows—projections of real movement, blue arrows—direction of perceived movement 47

Fig. 3.2 Typical daily coverage of the Earth by satellite measurements from a sun-synchronous orbit: areas of lack of data due to the discrepancy between successive scan swaths at the equator are

shown in gray, land in black. The example is built on the basis of data on the total precipitable water according to SSMIS measurements (DMSP *F*17) for 12/31/2015 (the color scale of TPW values in mm is shown below) 53
Fig. 3.3 Interpolation of SSM/I data on regular grid. Thick lines—grid cell of GRID (0.5°); circles indicate the measurements of SSM/I that fall into the GRID cell; thin lines visualize: **a** SSM/I scan lines, **b** 0.1° grid cells. 55
Fig. 3.4 The effect of measurement asynchrony during joint interpolation: **a** $\{s_1, s_2, s_3\} = \{F13, F14, F15\}$ DMSP satellite data; **b** $\{s_1, s_2\} = \{F14, F15\}$ DMSP satellite data. The color scale of the *W* values in *mm* is in the middle 58
Fig. 3.5 Lacunae stapling: **a** a fragment of the TPW field, coordinates in geographical degrees; black rectangle—the analyzed neighborhood of the boundary cell of the lacuna indicated by the black circle; gray rectangle—the block closest to the analyzed area according to the SAD metric and distant at a given number of cells from the lacuna; the white arrow is the calculated continuation of the field isoline at the boundary point (extrapolation direction); **b** the result of "stapling". The color scale of the field values in *mm* is below the figure 59
Fig. 3.6 Schematic illustration of the pyramidal method of motion estimation. The images on the left side are the total precipitable water values according to satellite radiometry data for June 8, 2016 (local time about 15:00); on the right side—for June 09, 2016 (about 03:00). The bottom row of images is analysis on the largest spatial scale. A displacement of the block bounded by a black frame by approximately 2.5° to the east was revealed. The top row of images is an analysis of the corresponding blocks on a more detailed spatial scale. The displacements of individual blocks are revealed, shown in the upper right image by arrows. Further, the analysis continues on successively decreasing spatial scales 64
Fig. 3.7 The average residual of the interpolated TPW fields according to SSMIS *F*16 with independent measurements of SSMIS *F*17 as a function of "synchronization" 69
Fig. 3.8 Distribution of interpolation residuals for cases **a** $\Delta t = +0.5$ h and **b** $\Delta t = -1$ h; approximation by the functions of Gauss (dotted line) and Cauchy-Lorentz (solid line) 69
Fig. 3.9 The average residual of the interpolated TPW fields according to SSMIS *F*16 with independent measurements of AMSR-2 as a function of "synchronization" 73

Fig. 3.10 The global TPW field on 08/28/2015 (color scale of TPW values below) at 12:00 UTC (**a**), 15:00 UTC (**b**), 18:00 UTC (**c**), 21:00 UTC (**d**) and the positions of the tropical cyclones Talim and Katrina (white circles and diamonds) at the corresponding time points . . . 75
Fig. 3.11 An iterative scheme for constructing and spatiotemporal interpolating of the fields of geophysical parameters of global coverage according to satellite radiometric observations . . . 77
Fig. 3.12 The TPW field the atmosphere over the Indian Ocean on 11/01/2013, reconstructed according to SSMIS *F*16, *F*17 and AMSR-2 without and with interpolation (see comments in text) . . . 79
Fig. 3.13 Scheme for calculating the flux through the contour element . . . 82
Fig. 4.1 Analysis of the evolution of TC Alberto: **a** the trajectory of the TC, color indicates different stages of development, geographical degrees are indicated along the edges of the figure; **b** the intensity of the TC (left scale) and the minimum pressure in the center of the eye (right scale), depending on the date in August 2000 (lower scale); **c** the TC in the TPW field and concentric circular contours of radii of 8 and 8.8° to calculate the power of the advective flux of latent heat; **d** the intensity of the TC (red line, left scale) and the power of latent heat fluxes (black lines, right scale) through predetermined contours, depending on the date in August 2000 (lower scale), see explanations in the text . . . 96
Fig. 4.2 TC Debby (North Atlantic): **a** TC trajectories and examples of integration contours; **b** the intensity of the TC (thick gray line, left scale) and the power of advective fluxes of latent heat (thin black lines, right scale), depending on the date in August 2000. See explanations in the text . . . 98
Fig. 4.3 TC Bilis (Northwest Pacific): **a** trajectory; **b** evolution . . . 99
Fig. 4.4 TC Ewiniar (Northwest Pacific): **a** the trajectory of the TC and concentric contours with radii of about 4° (thin black circles) and 8° (thick gray circles); **b** the intensity of the TC and the power of the advective fluxes of latent heat through the contours of 8°; **c** the same with 4° contours . . . 100
Fig. 4.5 TC Jelawat (Northwest Pacific): **a** trajectory; **b** evolution. See notes to Fig. 4.2 and explanations in the text . . . 101
Fig. 4.6 TC Giema (Northeast Pacific): **a** trajectory; **b** evolution . . . 102
Fig. 4.7 TC Hector (Northeast Pacific): **a** trajectory; **b** evolution. See notes to Fig. 4.2 and explanations in the text . . . 102

Fig. 4.8 Trajectories of the TC Haiyan (H) and TS Podul (P). Geographic coordinates are indicated in degrees north (N) and south (S) latitude and east (E) longitude 106
Fig. 4.9 **a** the intensity of the TC Haiyan (right scale) depending on the date in November 2013 at local noon (bottom scale) and the SST value (left scale) at the corresponding points of the TC trajectory (squares—one day before the TC passing; circles—one day after the TC passing; triangles—three days after the TC passing); **b** the same for the TS Podul 107
Fig. 4.10 **a** Evolution of Haiyan: a thick black line is the intensity of the TC (left scale), thin gray lines are the powers of advective fluxes of latent heat (right scale) through concentric circular contours with radii of 8 and 8.5° depending on the date in November 2013 at local noon (bottom scale); **b** the same for the TS Podul (radii of circular boundaries 4° and 4.5°, respectively); additionally indicated are the time intervals for the passage of the storm over land (Mindanao and Palawan Islands; Vietnam) 107
Fig. 4.11 **a** SST field in the North Atlantic (the color scale of temperature intervals with a step of 1 °C at the bottom of the figure), the trajectory of the TC Alberto (red line) and the position of the TC (red circle) on Aug 18, 2000; along the edges of the image—geographical coordinates; **b** plots of the TC intensity *V*, m/s (red line, left scale), average SST under the TC eye (green line, right inner scale) and the power of the advective flux of latent heat through the 8° boundary (black line, right outer scale); on the lower scale—dates in August 2000 108
Fig. 4.12 Evolution of the TCs Goni and Atsani: trajectories (black lines) and intensities (red lines, vertical scale). See explanations in the text 110
Fig. 4.13 The movement of the TCs Goni and Atsani relative to the geometric center of the system. The "+" symbols mark the beginning of the day (in UTC) on August 13–25, 2015, respectively. The phases of the TCs (TL, TD, TS, STS, T, L) are given according to (Pokrovskaya and Sharkov 2016) 111
Fig. 4.14 Scheme of the main tropospheric flows in the system of twin TCs Goni and Atsani: A—region of strong divergent flows branching to the centers of Goni (1) and Atsani (2); B is the region of weak unstable flows; 3-stream covering the TC system in the cyclonic direction. The background image is based on NOAA archival data (www.nnvl.noaa.gov/view/globaldata.html#TRUE) on August 18, 2015 112

Fig. 4.15 The time course of the intensity (m/s) of Goni, V_G (circles, a thin black line), Atsani, V_A (diagonal crosses, a thin black line), as well as the normalized characteristics of the total intensity of the system of the two TCs: linear, V_1 (thick black line), and quadratic, V_2 (thick red line) 113
Fig. 4.16 Goni and Atsani in the TPW field (color scale on the right) and the families of contours considered: 1—around the TC Goni with r = 3°, 6°, 8°; 2—around the TC Atsani with r = 2°, 4°, 6°, 8°; 3—composite with R = 9°, 11° 114
Fig. 4.17 Evolution of the Goni and Atsani system in comparison with latent heat advection: **a** Goni intensity, V_G (line 1, left scale) and latent heat advection, Q (lines 2–4, right scale) through the contours around the TC with radii of 3°, 6°, 8°, respectively; **b** Atsani intensity, V_A (line 1, the left scale) and advection of latent heat, Q (lines 2–5, the right scale) through the contours around the TC with radii of 2°, 4°, 6°, 8°, respectively; **c** the intensity of the Goni and Atsani system, V_2 (line 1, left scale) and the latent heat advection, Q (lines 2–4, right scale) through the composite contours around the system (see the example of circuits in Fig. 4.16), at R = 11°, 9°, 8° 115
Fig. 5.1 A fragment of the TPW field over the Pacific Ocean (color scale in mm on the right) for December 1, 2016: **a** mosaic according to SSMIS F16 and SSMIS F17; **b** a product of satellite radiothermovision. At the edges, the geographical coordinates of the fragment in degrees; positive values for the northern and eastern hemispheres, negative values for the southern and western 123
Fig. 5.2 The TPW field over the northeast Pacific Ocean and the North Atlantic for 03/11/2016 using the satellite radiothermovision algorithm (color scale and coordinate notation—as in Fig. 5.1). TPW values above the following thresholds are shown: **a** 20 mm; **b** 23 mm; **c** 26 mm; **d** 30 mm. The scale sections in the upper figures correspond to 1000 km 125
Fig. 5.3 The TPW field over the North Atlantic with superimposed elements of the advection field; calculation on the dates of observations: **a** 03/21/2016; **b** 06/01/2016. The color scale and designations of coordinates—as in Fig. 5.1. 126
Fig. 5.4 The TPW field (color scale—see Fig. 5.1), recalculated to 12/02/2016 01:00 UTC (**a**), and the difference of the fields shown in Figs. 5.1b and 5.4a (**b**), the color scale—to the right 127
Fig. 5.5 Schematic flowchart of automatic detection of atmospheric rivers in the total precipitable water field of the atmosphere. . . . 129

Fig. 5.6 **a** Selection of a fragment of the TPW field for a given basin (Northern Hemisphere, Pacific Ocean), the color scale of the TPW values and designation of coordinates—as in Fig. 5.1; **b** refinement of the latitudinal border in the tropics (the mask of the areas excluded from the analysis is shown in purple); **c** the allocation of air masses of mid-latitudes (the mask of the areas excluded from the analysis is shown in purple) 130
Fig. 5.7 Analysis of the histogram of TPW values of the selected field fragment (see Fig. 5.6b). On a horizontal scale—TPW values, on a vertical—the number of corresponding TPW values. Thin gray line is the initial histogram; the red line is a smoothed histogram; light green line—approximation of the first peak; the blue line is the residue from the subtraction from the smoothed histogram of the first approximated peak; cyan line—approximation of the second peak; the pink line is the residue from the subtraction of the second approximated peak; dark green line—approximation of the fourth peak; the yellow–green line is the remainder of the subtraction of the fourth peak. Designations correspond to Eqs. (5.2–5.11). See explanations in the text 132
Fig. 5.8 Block diagram of the morphological analysis of a fragment of the TPW field 135
Fig. 5.9 Smoothing the TPW field by median filtering: **a** a fragment of the initial TPW field; **b** the same fragment after smoothing. The color scale of the TPW values and the notation of coordinates—as in Fig. 5.1 136
Fig. 5.10 Marking the slopes and plateau of the smoothed TPW field: **a** a fragment of the smoothed TPW field (as in Fig. 5.9b); **b** a map of slopes and plateaus: the northern slopes are marked in blue, the southern slopes are orange, the plateau is yellow–green, see the text for explanations 136
Fig. 5.11 The procedure for extending slopes: **a** the initial map of marked slopes and plateaus (as in Fig. 5.10b); **b** the result of the slope extension procedure, see the explanations in the text 137
Fig. 5.12 Localization of ridges: cells corresponding to ridges according to criterion (5.14) are marked in white in the figure; the background is the same as in Fig. 5.11b (on an enlarged scale) 137
Fig. 5.13 The result of combining ridges 139
Fig. 5.14 **a** Longitudinally-oriented and **b** latitudinally-oriented ridges in the TPW field; **c** the result of their combination (the restored fragments of the latitudinally oriented ridges are shown in blue; the longitudinally oriented ridges in white) 139

Fig. 5.15 The detected axes of atmospheric rivers (black lines) superimposed on the global field of the TPW of the atmosphere over the ocean (color scale—as in Fig. 5.1). The land mask is shown in gray. At the edges of the picture—geographical coordinates in degrees 141
Fig. 5.16 Advection of latent heat in an atmospheric river: **a** an atmospheric river in a TPW field above the North Atlantic on 10/21/2016, a color scale of TPW values on the right; **b** a graph of the power of the advective flux of latent heat across the 45° N boundary as a function of longitude superimposed on the corresponding fragment of the TPW field (marked with a black frame in Fig. 5.16a). See explanations in the text 143
Fig. 5.17 Phases of evolution of atmospheric rivers in the TPW field over the North Atlantic: **a–f** 01/01/2016; **g** 01/03/2016; **h** 01/10/2016. The color scale of TPW values as in Fig. 5.1 . . . 145
Fig. 5.18 Global fields of total precipitable water Q, mm (**a**); cloud liquid water content L, mm (**b**); surface wind speed W, m/s (**c**) and their product P, rel. units (**d**) based on satellite radiometry remote data for December 25, 2016. At the edges of the images—geographical coordinates in degrees. Color scales of field values are to the right of the corresponding images 147
Fig. 5.19 Global TPW field over the ocean and land according to the SSM/I, SSMIS, WindSat and ASMR-2 data for June 28, 2013, combined using satellite radiothermovision algorithms. The color scale of TPW values is as in Fig. 5.1. At the edges of the image, the geographical coordinates in degrees. The boundaries of the continents and areas of data gaps are indicated by a purple mask 148
Fig. 6.1 Combined fields of total precipitable water (color scale in mm at the bottom) and advection rates (vectors, calibration standards on the right): at the top—calculation according to data on 01/01/2003; below—on 10/01/2017. At the edges of the images, the latitudes and longitudes in degrees 154
Fig. 6.2 Border families for calculating latent heat fluxes: black—along parallels with a step of 5°; colored—along the meridians. The meridional boundaries marked with the letters "I" and "A" are used in the calculations illustrated in Fig. 6.6. 155
Fig. 6.3 Average field of integral atmospheric moisture content over the entire observation period, excluding incomplete 2017 (color scale in mm below) and the corresponding average field of advection velocities (vectors). On the right are the zonal components of the advection velocity v, m/s as a function of latitude θ along the meridional Section 1: 65° E; 2: 178° W; 3: 30° W in the range of 70° S up to 70° N. 156

Fig. 6.4 Seasonal average fields of integral moisture content and advection rates: **a** summer 2003; **b** winter 2003/2004 (see explanations in the text) 158
Fig. 6.5 Analysis of meridional fluxes of latent heat over the oceans: at the latitude of the equator (left); at 25° north latitude (right): time series of the total power *P*, PW of the flows with a sampling of 6 h (black curves) and three-month averaging (red curves); wavelet spectra (**a**, **b** above) and Fourier spectra (**c**, **d**) of the initial time series (explanations in the text) 159
Fig. 6.6 Analysis of zonal fluxes of latent heat in the tropics over the Indian (left) and Atlantic (right) oceans: time series of the total power *P*, PW of flows with a sampling of 6 h (black curves) and three-month averaging (red curves); wavelet spectra (**a**, **b** above) and Fourier spectra (**c**, **d**) of the initial time series (explanations in the text) 161
Fig. 6.7 Specific power of meridional latent heat fluxes as a function of latitude: blue curves—average "winter" flows, red curves—average "summer" flows; dashed blue curve—average flow of winter 2016/2017; dashed black curve—average flow of summer 2017; the thick blue curve is the average "winter" stream for all years, the thick black curve is the average "summer" stream; thick green curve—average flux for the continuous observation interval 2003–2016; triangles—recalculation of data (Palmén and Newton 1969) (explanations in the text) 163
Fig. 6.8 Specific power (MW/m) of average annual latent heat fluxes over the World Ocean through latitudinal boundaries from 60° S up to 60° N in increments of 5° for 2003–2016. Positive latitudes (along the abscissa) are for the northern hemisphere. Positive values of flows (along the ordinate axis)—for the direction to the north. The gray bar is the range of specific power values (for a given latitude) that differ from the average for all years by no more than one standard deviation 165
Fig. 6.9 Specific power (MW/m) of average annual latent heat fluxes over the Pacific Ocean through latitudinal boundaries from 60° S up to 60° N in increments of 5° for 2003–2016. Positive latitudes (along the abscissa) are for the northern hemisphere. Positive values of flows (along the ordinate axis)—for the direction to the north. The gray bar is the range of specific power values (for a given latitude) that differ from the average for all years by no more than one standard deviation 167

Fig. 6.10 Specific power (MW/m) of average annual latent heat fluxes over the Atlantic Ocean through latitudinal boundaries from 60° S up to 60° N in increments of 5° for 2003–2016, see the notes to Fig. 6.9 169
Fig. 6.11 Specific power (MW/m) of average annual latent heat fluxes over the Indian Ocean through latitudinal boundaries from 60° S up to 20° N in increments of 5° for 2003–2016, see the notes to Fig. 6.9 171
Fig. 7.1 The basic architectural diagram of the geoportal; the transition between data levels corresponding to external sources, geoportal servers and user environment is carried out using specialized geoportal procedures 178
Fig. 7.2 Diagram of the use of RSS products in the construction of the reference collection of the geoportal: products marked with shaded bands are used on a regular basis; vertical numbered lines show the intervals for attracting additional data (white bars), see Table 7.1 179
Fig. 7.3 The order page for custom products (interpolated geophysical atmospheric fields over the World Ocean) of the geoportal of satellite radiothermovision 182
Fig. 7.4 Visualization of the result of the calculation **a** by the Eq. (7.1); **b** by the Eq. (7.2) 185
Fig. 7.5 Visualization of the calculation (7.5) for the dates: **a** August 28, 2013; **b** February 28, 2013 186
Fig. 7.6 Syntactic graph for the concept of *equation* in ICAR 187

List of Tables

Table 2.1 Requirements for the parameters of satellite measurements of the atmospheric total precipitable water according to WMO OSCAR 30
Table 2.2 Some satellite microwave imagers/sounders operating in 2000–2020 33
Table 3.1 Distribution characteristics of interpolation residuals 70
Table 3.2 Parameters for approximating the distribution of residuals by the Cauchy-Lorentz function 72
Table 4.1 Tropical cyclones of August 2000 95
Table 6.1 Meridional circulation of latent heat over the World Ocean 166
Table 6.2 Meridional circulation of latent heat over the Pacific Ocean . . . 168
Table 6.3 Meridional circulation of latent heat over the Atlantic Ocean 170
Table 6.4 Meridional circulation of latent heat over the Indian Ocean 171
Table 7.1 Supplementing main data at problematic intervals shown in Fig. 7.2 180

Chapter 1
Introduction

This brief introductory chapter is addressed mainly to readers who are just beginning their acquaintance with satellite radiometric remote sensing of the Earth, and unlike the rest of the chapters, it is intentionally written in a popular form. Although it does not contain facts that would not be well known or obvious to experts, the chapter may be still useful to them. It explains the general motivation and logic of the research that led to the creation of the satellite radiothermovision approach of atmospheric processes described in the book.

1.1 Weather, Climate and the Water Vapor

The Earth's climate system is in many ways a unique physical object. And above all, in that this is our home—the only habitat that has been so far mastered by mankind to some extent. This explains our desire to have as much detailed and accurate knowledge about it as possible. At the same time, this explains why the main method of scientific knowledge, which has proven itself in most other cases, is not suitable for this purpose. It would be too irresponsible to set up a real physical experiment over the Earth's climatic system, even if mankind had the energy resources that would make it possible to direct and control this experiment.

The safe way for the study of the climate system, found by scientists, consists in mathematical modeling—the study of hypothetical scenarios of its development in various conditions. Short-term detailed scenarios form the basis of numerical weather forecasts. Long-term global scenarios are needed to analyze and predict climate change. A grounded construction of such climatic scenarios (especially long-term ones) should be based on the foundation of knowledge accumulated by mankind in almost all natural science disciplines—from astronomy and physics of the Sun to hydro-dynamics and thermo-dynamics of natural environments, from studies of ionospheric disturbances to the physics and chemistry of processes that occur deep

D. M. Ermakov, *Satellite Radiothermovision of Atmospheric Processes*, Springer Praxis Books,
https://doi.org/10.1007/978-3-030-57085-9_1

under the earth's crust, from microprocesses of the exchange of matter and energy in the biosphere to planetary geodynamics and processes of landscape formation. The so-called anthropogenic factor has to be singled out as a special element of the climate system.

It is important to realize that even an accurate knowledge of the laws of the development of the system alone will not tell what this system will be in the future. At a minimum, it is still necessary to characterize its current state, the starting point of development. We can get this information only from the data of actual observations. And if one receives such information regularly—that is, *monitor* the necessary parameters, then it's possible, among other things, to compare the accumulated results of observations with the forecast scenario and to make the necessary corrections in the model concepts. Generally speaking, the more reliable and detailed weather and climate forecasts we want to achieve, the more complete and more accurate should be our observations of an ever-changing climate system.

The most dynamic and variable component of the Earth's climate system is its atmosphere. In turn, one of the most important gas components of the Earth's atmosphere, which has a strong influence on weather and climate processes, is water vapor. Water is not only a cradle and a necessary element of earthly life. In the Earth's climate system, it plays an important role as an energy agent that helps maintain relatively comfortable, familiar (to us) conditions. In this narrow range of conditions, water, unlike many other substances, is able to be both in a solid, in a liquid, and in a gaseous state. In middle latitudes, for example, this can often be observed on a rainy spring day, when atmospheric water vapor condenses and falls in the form of rain on the ground, still covered with snow or ice somewhere.

Transitions of matter from one state of aggregation to another are accompanied by the release or absorption of energy. So, water vapor, forming over a heated tropical ocean, takes away the energy of solar radiation expended on its formation. Moving with air currents, water vapor redistributes this energy (the so-called *latent heat*) in the Earth's atmosphere. The latent heat will be returned into the atmosphere in the reverse process of condensation of water vapor. Most of the released energy will go into the chaotic movement of molecules (heating), but a certain fraction (usually no more than one percent) can turn into an ordered, directed movement—the kinetic energy of the air.

Therefore, an abrupt formation of heavy precipitation can lead to storms, rapid intensification of tropical cyclones to the stage of a hurricane (typhoon) and other catastrophic atmospheric events. Extreme weather conditions on the coasts (and sometimes deep in the continents) often occur during the land-falls of tropical moist air masses (in particular, the so-called *atmospheric rivers*).

On the other hand, the gradual "spreading" of atmospheric moisture from the equator to the poles (*meridional transport*) leads to additional heating of high latitudes due to the tropics. The water vapor content in the air is relatively small: on average, the total mass of water vapor is estimated at about 0.2% of the mass of the "dry" atmosphere (1.27×10^{16} kg vs. 5.14×10^{18} kg). Nevertheless, the meridional flow of latent heat created by water vapor is comparable with the heat flux created by the rest of the warm air, as well as the heat flux created by the waters of the ocean.

The role of these three factors in the planetary energy balance is easy to evaluate by comparing the "mild" climate of the Arctic, open to the influx of all three types of heat, with the harsh conditions of the Antarctic. The mainland of Antarctica, covered with a thick ice sheet, and the closed (*circumpolar*) currents formed around it in the ocean and atmosphere significantly impede the flow of heat to the south pole.

1.2 Satellite Microwave Radiometry of the Atmosphere

The physical opportunity to "take a look" at the Earth's atmosphere as a whole arose only in the second half of the last century with the beginning of the era of space exploration. However, the capabilities of human vision, as well as any means of observation in the visible spectrum, is clearly not enough for a detailed study of the atmosphere (and surface) of the Earth. At any given time, half of the Earth is immersed in the night. The opposite day side, in turn, is always about half-covered by clouds that do not allow light to pass through. Therefore, one of the most effective ways to study the atmosphere at any time of the day and in any weather conditions was the observation of its own thermal radiation in the microwave range (at wavelengths of the order of 10^{-3}–10^{-1} m). Almost half a century has passed since the launch of the first satellite microwave radiometer to observe the Earth. During this time, satellite microwave radiometry has impressively expanded its capabilities, the completeness and reliability of the information received with its help has increased many times over.

Figure 1.1 illustrates the distribution of water vapor in the Earth's atmosphere on September 2, 2019, retrieved from observations of the SSMIS instrument onboard the DMSP F17 satellite (this and other satellite radiometers are described in Sects. 2.2 and 2.3 of Chap. 2). Using this figure as an example, we will examine in more detail some of the capabilities of modern satellite microwave radiometry.

1.2.1 What *Do We See?*

To begin with, we will figure out what physical quantity (the geophysical parameter of the atmosphere) is depicted in the Fig. 1.1. In the scientific literature, it has several names, including "columnar water vapor", "integrated water vapor", "total column water vapor", etc. This book uses the term "total precipitable water" and its abbreviation TPW. TPW characterizes the amount of water vapor in a vertical atmospheric column. Imagine that we have chosen a certain area on the surface of the Earth and surrounded it with infinitely high vertical walls. If all the water vapor trapped inside these walls is condensed, a layer of water will appear on the surface. The thickness of this layer (usually measured in millimeters) is the value of TPW. (How TPW values are retrieved from radiometric observations and what other

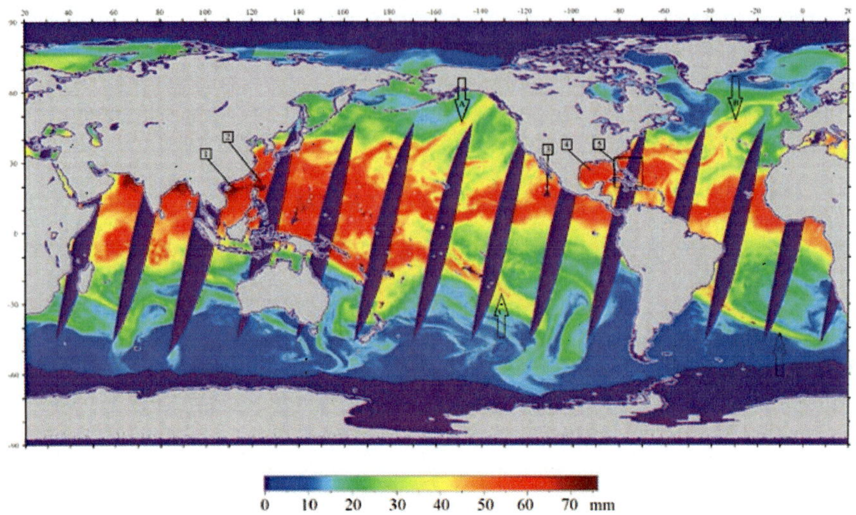

Fig. 1.1 Distribution of the total precipitable water in the Earth's atmosphere on September 2, 2019, the color scale of the values below; gray color—land; purple—data gaps (see explanations in the text)

geophysical parameters can be obtained from them is briefly described in Sect. 2.2 of Chap. 2).

Pay attention to the color scale of TPW values under the figure. The choice of range suggests that the atmospheric TPW is expected to rarely exceed 75 mm. The highest values are naturally observed in the tropical zone, where the evaporation of water from the ocean heated by the Sun is most intense. Here they often make up more than 50 mm. Outside the tropics, TPW generally does not exceed 30 mm. Thus, there is really "very little" moisture in the atmosphere: compare the layer thickness of 50 mm (0.05 m) with the average ocean depth (about 3700 m)!

A closer look at Fig. 1.1 shows that the real picture of the distribution of water vapor in the atmosphere is much more complicated than the draft scheme described above. Water vapor is an indicator of atmospheric motion. Moving together with air currents, it "visualizes" atmospheric processes at different stages of their development. So, for example, tropical cyclones, swirling air flows, draw together the water vapor in the surrounding atmosphere to the center of rotation. As a result, tropical cyclones are distinguished by increased TPW values, forming large structures in the shape of a spiral or comma.

In Fig. 1.1, at least four such structures are distinguished, indicated by numbers in rectangular frames. This, from left to right, is the tropical storm Kajiki (1) and typhoon Lingling (2) in the northwest Pacific Ocean, major hurricane Julliette (3) in the northeast Pacific Ocean and the strong tropical storm Fernand (4) over the Gulf of Mexico in Atlantic Ocean. (The categories "tropical storm", "typhoon", "hurricane" etc. are used to classify tropical cyclones by intensity and have regional features. For more details on the energy characteristics of tropical cyclones and the classification system for their intensity adopted in this book, see Chap. 4).

Of course, as an indicator of atmospheric movements, in principle, any substance carried away by air currents is suitable if the change in the content of this substance in the atmosphere can be reliably measured from a satellite. Water vapor is unique in that, in addition, it does not remain a passive participant in ongoing events. On the contrary, it plays the role of the energy element of atmospheric processes, storing solar heat during evaporation and releasing it during condensation. How significant is this role? Note that each of the tropical cyclones indicated in Fig. 1.1 can be enclosed by a circular contour of a diameter of about 10 geographical degrees (more than 1000 km at the equator), inside which TPW values are 50 mm and higher. It is not difficult to evaluate the "latent heat" stored in this mass of vapor. It is equal to the total condensation energy that will be released if all the water vapor—and this is 50 kg per every square meter—simultaneously turns into a liquid inside a circle with a diameter of 1000 km $= 10^6$ m (i.e., an area of about 0.75×10^{12} m^2). Assuming the specific heat of vaporization and condensation of water to be 2.26 MJ/kg, we get 8.5×10^{19} J as a result. What is this value compared with? For example, at the current rates of world electricity production, estimated in order of magnitude of 0.01 PW $= 10^{13}$ J/s, the generation of such an amount of energy requires almost 100 days of continuous operation. Recall that we are talking only about one tropical cyclone!

1.2.2 Where *Do We See?*

Since it is already clear that the picture of the distribution of water vapor in the atmosphere is so important for studying the processes occurring in it, the next important question is why is it so fragmented in Fig. 1.1? First of all, there is no information on TPW values over land, as if observations were made only over the ocean. This is certainly not the case. But due to the variety of properties of the land itself, the interpretation of radiometric measurements above it is in most cases too complicated to provide reliable TPW retrievals. Only a few modern satellite devices, due to their design features, make it possible to obtain some estimates, although much less reliable than over the ocean. (This issue is addressed in Sect. 2.2 of Chap. 2). It is important to note here that both air humidity and wind speed on average reach much higher values over the ocean than over land. Therefore, it is precisely the pattern of water vapor distribution over the ocean that is most indicative of the study of many atmospheric processes and the climate system as a whole. At the same time, the development trends of satellite radiometry allow us to hope that already in the next decade we will have sufficiently reliable means of TPW retrieval over all types of the Earth's surface.

In addition to the lack of information about TPW over land, we note in Fig. 1.1 that the observations are grouped in stripes separated by significant gaps, especially wide at the equator. As a result, even over the ocean there are areas (called "lacunae" in the book) in which data are not available. What is their nature, and what is the pattern of their appearance?

In short, the reason is that it is necessary to clearly distinguish relatively small details in the overall picture of the distribution of TPW.

It would seem that when observing from space it is logical to rise to a great height to simultaneously observe almost the entire disk of the Earth. That is how geostationary meteorological satellites function, "hanging" at an altitude of about 36,000 km (which is about 5.5 times the radius of the Earth) above a given equator point and every half hour, fifteen minutes, and sometimes more often, transmitting images of the planet in the visible and infrared spectra. Unfortunately, as you move away from the object of observation, its outlines lose their distinctiveness. The minimum size of confidently distinguishable details of an object, which determines the spatial resolution of observations, is proportional to the distance to it. So, at the height of the geostationary orbit, observations in visible light have a spatial resolution of about 1 km and make it possible to clearly distinguish large cloud systems and monitor the development of the so-called mesoscale and synoptic atmospheric processes (Sect. 2.1 of Chap. 2 briefly talks about them).

The problem with microwave radiometric measurements, however, is that the minimum size of distinguishable details is also proportional to the observation wavelength. And while the middle of the visible spectrum corresponds to a wavelength of about 550 nm or 5.5×10^{-7} m, the maximum of radiothermal radiation of a water vapor molecule falls at a wavelength of 1.35 cm or 1.35×10^{-2} m, which is approximately 25000 times more. With a spatial resolution of even 2500 km, the largest typhoons would look like faint points, devoid of any internal structure.

For this reason, microwave radiometers are located in much lower orbits—with a height of 700–800 km above the Earth. (This is 45–50 times lower than the geostationary orbit, but still twice as high as the orbit of the International Space Station). Even with such a "zooming", a portion of the Earth's surface measuring several tens of kilometers merges into a single point for a satellite radiometer, but such spatial resolution is already sufficient to reconstruct the patters of TPW as in Fig. 1.1. However, the method of forming such a picture is completely different than when observed from geostationary satellites.

To stay in orbit at an altitude of 800 km, satellites must move at a speed of about 7.5 km/s, making a complete revolution around the Earth in about an hour and a half. From this altitude, only an area of 1500–2000 km falls into the field of view of the satellite radiometer, therefore, to cover the entire Earth with observations, it is necessary to fly around not in the equatorial plane, but in a highly inclined orbit—from pole to pole. With this movement, observations form stripes (*swaths*) up to 2000 km wide, which "wrap" around the Earth's globe like a thread on a ball. (We emphasize that the width of the swaths remains constant everywhere, and their apparent broadening at high latitudes in Fig. 1.1 is caused by projective distortions, see also Fig. 3.2 in Chap. 3). During one loop, corresponding to the period of the satellite's revolution, the Earth manages to turn slightly eastward, and each subsequent swath passes west of the previous one. For a day, the Earth makes a complete revolution around its axis, and the satellite observes it from all sides.

Continuing the analogy with a ball and thread, it is easy to understand that near the poles many swaths overlap each other, and therefore the satellite observes regions of high latitudes more often than others. Unfortunately, the opposite situation arises near the equator: the distance between adjacent swaths is greater than their width, the swaths diverge, and those "lacunae" appear that are clearly visible in Fig. 1.1.

1.2.3 When *Do We See?*

Among all the low orbits with a high inclination, the so-called *sun-synchronous* orbits occupy a special place. Their parameters are selected so that the satellite always flies over the equator at the same local time. This is convenient for two reasons. Firstly, this allows more effectively taking into account and minimizing the effect of direct and reflected from the Earth solar radiation on the device: on each pass of the orbit, the Sun repeats the same path through the sky relative to the satellite. Secondly, the parameters of the atmosphere and surface (for example, their thermodynamic temperature) can substantially depend on the time of day, and synchronization with the Sun allows to bring the conditions of observations closer to each other and more stable.

Of course, it should be borne in mind that Fig. 1.1 obtained as a result of such observations does not reflect the instantaneous state of the atmosphere, but rather shows a time-based sweep of this state. The swath crossing the Bay of Bengal (in a southerly direction) corresponds to the beginning of September 2, 2019, according to the universal coordinated time. Further, the swaths with a period of about 1.5 h sequentially shift to the west, to the eastern coast of Africa, reappear at its western coast and continue to shift westward. The last swath for the current day passes to the west of Australia. It is seen (especially in the northern latitudes) that the spatial structure of the TPW field experiences sharp discontinuities at the boundary between this swath and the one to the left of it. This is natural, since the corresponding observations are no longer divided by an hour and a half, but by a daily interval.

In addition, it is noticeable that the lacuna separating the first and last swaths in the current day is slightly wider than the other ones. This is because the length of the day is not exactly a multiple of the satellite's orbital period. The swaths performed on the next day will be slightly shifted eastward relative to those observed in Fig. 1.1. As a result, the lacunae separating them will also be shifted. Thus, the satellite does not have permanent "blind spots"—each point on the Earth is guaranteed to fall into the field of view within several days.

Another important point to make. In fact, Fig. 1.1 illustrates exactly half the daily coverage that a satellite device forms in a sun-synchronous orbit. How does the other half come about?

Figure 1.1 corresponds to the observations in the so-called *descending* parts of the satellite orbit. Each swath shown in the figure begins near the north pole and crosses the Earth in a southerly direction, following the movement of the satellite. (This part is called descending because the satellite moves from north to south, that is, in the traditional European view from top to bottom). Having circled the Earth near the south pole, the satellite continues to observe—now in the direction from south to north, that is, in the *ascending* part of the orbit. Remembering the analogy with a ball and a thread, it is easy to understand that on the descending and ascending parts of the orbit, the satellite crosses almost diametrically opposite points of the equator, and the local time at these points differs by half a day. That is how much time it takes the Earth to make half a turn around its axis when these points "change places"—the first appears in the ascending part of the orbit, while the second is in the descending part.

Thus, the satellite forms two diurnal patterns similar to those shown in Fig. 1.1, but with different slopes of the swaths and with different (by 12 h) local times of observations. So, for example, Fig. 1.1 is constructed according to the morning observations (the equator crosses around 6:30 local time) in the descending parts of the orbit. In addition to it, there are evening observations (the equator crosses around 18:30 local time) in the ascending parts of the orbit. Thus, distracting from the details, we can say that satellite devices in solar-synchronous orbits observe the entire planet twice a day.

The noted features show that in the interpretation of satellite radiometry data (especially in the analysis of dynamic atmospheric processes), the use of analogies with the analysis of static images requires certain caution and, sometimes, special processing techniques.

1.2.4 What **Don't** *We See?*

The answer to this question seems simple. We do not see all those regions of space at all those moments in time when a satellite does not pass over them. For example, we do not see how the TPW field is arranged inside the lacunae. A simple solution begs to launch a second satellite device, which will pass over the lacunae at the same local time.

As we remember, this approach will allow us to obtain a more or less complete picture of the state of the atmosphere only twice a day. However, given how fast atmospheric processes sometimes develop and what danger they can pose to humans, we would probably like to increase the frequency of observations. Not for nothing meteorological geostationary satellites record the state of the atmosphere at least once every half hour. Suppose we want to reduce the period of a full Earth survey by satellite radiometers from 12 to 1.5 h. So, instead of two identical instruments operating in orbit at the same time, it is already necessary to have 16 of them. And to guarantee a smooth replacement in the event of failure or temporary malfunctions of one instrument, it is advisable to keep a few more standby instances in orbit.

Without going into details, it can be stated that at the moment such a solution still seems fantastic. And certainly, it has never been realized in all the past time of observations that recorded the most valuable factual information about decades of evolution of the Earth's climate.

Is it possible to draw up a complete picture with fewer satellites—preferably just those at our disposal? Indeed, at first glance it seems understandable how the TPW field is arranged inside the lacunae—the eye itself prompts the most logical solution, smoothly connecting and "pulling together" the borders of neighboring swaths. Similarly, studying the chronological sequence of observations in increments of 12 h, we can conclude that the TPW fields change in time smoothly and naturally, and basically as "the eye suggests".

Indeed, it turns out to be possible to construct a formalized analogue of this intuitive solution (the specific method is described in Chap. 3 of this book) and prove, by comparison with independent observations, that it is characterized by an acceptably high accuracy. However, the data of one satellite is still obviously not enough for this. A simple but convincing illustration is again given in Fig. 1.1. In it, just east of the Florida peninsula, a rectangular area is marked, indicated by the number "5". Approximately half of the points in this region fell into the lacuna between the swaths or are near land, so the TPW values in them are unknown. At the edges of the region, the TPW values are distributed in such a way that the eye suggests a rather trivial, "even" field structure inside the gap without any features. Reality is just the opposite. In the marked area, another tropical cyclone is hiding—major hurricane Dorian, which still maintains the highest intensity category on September 2 (although the intensification peak was reached the day before, on September 1) with a maximum wind speed of 155 knots (about 80 m/s) and a minimum pressure in its center of 914 mbar. Thus, the analysis of atmospheric dynamics requires a certain "information redundancy", which is provided by several satellite radiometers operating simultaneously in orbits. The consideration of evolution of tropical cyclones in TPW fields (as well as some other geophysical fields) is the subject of Chap. 4 of the book.

1.2.5 A Broader View in Conclusion

In conclusion, let us touch upon another important aspect. As we recall, Fig. 1.1 reflects the global state of the Earth's atmosphere. Nevertheless, contrary to expectations, it, at first glance, does not reflect the most fundamental structure of the atmosphere—namely, the global circulation cells. The existence of these cells belongs to the most general, basic knowledge about the atmosphere. As we know, atmospheric air, warming up especially much near the equator, becomes lighter and rises up due to convective forces. In high rarefied layers, it gradually loses its buoyancy and spreads in the poleward directions, gradually cooling. As a result, its density increases, and it descends to the surface of the Earth, but at a considerable distance from the equator. From here, it is again drawn to the equator due to the zone of reduced pressure formed there by the ascending flows (the so-called *intertropical convergence zone*). As a result, tropical air circulates in both hemispheres through giant closed loops, called the Hadley cells, in honor of the researcher who first described this mechanism. At mid-latitudes, a similar, but oppositely directed, Ferrel cell forms. At high latitudes, polar circulation cells exist in the least pronounced form. It is believed that the boundaries separating the Hadley and Ferrel cells are located at approximately 30° latitude in both hemispheres.

The mechanism of formation of global circulation cells is described so simply and clearly that there appears no doubt about the existence of real physical boundaries, which if not exclude, then in any case greatly impede the mutual penetration of tropical and polar air masses, preventing the exchange of matter and energy between them. However, daily observations of atmospheric dynamics (in which it is difficult to overestimate the role of satellite data) would seem to indicate the opposite. Thus, tropical cyclones, originating near the intertropical convergence zone, regularly rise to middle and high latitudes. The major hurricane Dorian mentioned above during its existence caused weather disasters and damage not only on the southeastern coast of the United States, but also on the northeastern coast of Canada. Its trajectory began from 10° north latitude at the time of its generation on August 24 and crossed 51° north latitude just before the final filling on September 9, 2019.

Attention is also drawn to phenomena of substantially greater spatial scales, for example, narrow bands of high TPW values that periodically form in the atmosphere and stretch for several thousand kilometers from the intertropical convergence zone to high latitudes. Four examples of such bands (two in each hemisphere) are indicated in Fig. 1.1 by vertical arrows marked “A”, “B”, “C”, “D”. Observation and modeling data indicate that at least some of these bands are peculiar channels for the rapid transfer of atmospheric water vapor from the tropics to middle and high latitudes—the so-called atmospheric rivers, which are discussed in Chap. 5.

How do the observed facts and the model of global circulation cells agree with each other? The approach to the analysis of atmospheric dynamics described in the book allows answering this question as follows. The idea of circulation cells is valid only when averaging data at the largest spatiotemporal scales. The circulation of air masses in the Hadley and Ferrel cells can be considered as a background process that is perturbed and “deformed” by other processes of various scales—including the planetary ones. The characteristics of this “background” process can be restored by accumulating and analyzing data from long-term observation intervals. On the whole, they are in good agreement with the general theoretical concepts of global atmospheric circulation, but they open up new opportunities for a more detailed analysis of trends related to climatic changes in recent decades (in particular, the so-called *widening of the tropics*). Chapter 6 of the book is fully devoted to these questions.

It should be noted that the results described in Chap. 6 are of interest in another important respect. They indicate that many key parameters of the dynamics of the Earth’s atmosphere (daily and seasonal periodicity, the boundaries of the circulation cells, the direction and speed of the zonal transport in them, the total transport of water vapor from the southern hemisphere to the northern one, etc.) can be restored exclusively from satellite radiometric observations, without involving additional data and a priori knowledge about the Earth’s climate system. Hence, this information is indeed contained in the observational data themselves, it can be extracted from them, for example, by the method described in the book and used as independent to verify and refine model representations.

Of course, the author is aware that any scientific research (and his own for sure) cannot claim to receive final and absolutely true results. Work on the development and improvement of the method of satellite radiothermovision continues. It is impossible to include in the book a set of new data that is still being prepared for publication. It is quite possible and even expected that in the process of their more detailed analysis, some ideas and conclusions formulated by the author and his colleagues will be subjected to refinement and development. The author sees the main task of the book not in fixing some particular results, but rather in attracting the attention of the scientific community to new opportunities for analyzing satellite information based on the proposed approach, and will consider its goal to be fully achieved if the book gives rise to active independent research in this direction.

To partly facilitate such a research the author and his colleagues have made significant efforts to make products of satellite data processing by their algorithms freely available to the scientific community through the geoportal of satellite radiothermovision. Its description is given in Chap. 7 of the book.

Chapter 2
Why Satellite Radiothermovision?

This chapter reveals the prerequisites for the creation of the method of satellite radiothermovision of mesoscale and synoptic atmospheric processes described in the book as an approach to solving specific inverse problems of remote sensing of the Earth by radiophysical means. For this purpose, the first section gives a brief description of the objects of study. It is shown that in the study of the formation and (or) evolution of the atmospheric processes under consideration, the main, and in some cases, the determining role, requiring careful quantitative analysis, is played by advection (horizontal movement) of water vapor in the lower troposphere. Obtaining sufficient volumes of reliable information about the moisture content of the lower troposphere is provided, first of all, by satellite radiothermal (passive microwave) observations. The second section of the chapter is devoted to a brief overview of the history and current state of development of these observational means. The short but important third section is devoted to satellite radiometric instruments (imagers and sounders) that are functioning in orbit now, or were operational in the nearest past. The latter is especially significant for constructing multiyear datasets to collect representative statistics on various types of atmospheric processes or the global atmospheric circulation as a whole. Finally, the fourth section briefly describes the idea of "dynamical analysis" of geophysical fields put in the core of the satellite radiothermovision approach. This idea appears to be very fruitful and applicable in a wide range of Earth remote sensing problems. The key aspect focused in this book is the retrieval of advection rates of water vapor in the lower troposphere, and, consequently, one of the most important atmospheric energy characteristics—vertically integrated horizontal fluxes of latent heat (potential energy of phase transitions of atmospheric moisture) through predetermined boundaries.

D. M. Ermakov, *Satellite Radiothermovision of Atmospheric Processes*, Springer Praxis Books,
https://doi.org/10.1007/978-3-030-57085-9_2

2.1 Elements of Mesoscale and Synoptic Atmospheric Dynamics

This section gives a preliminary description of the objects of study and substantiates the choice of specific atmospheric phenomena considered in the book. Later we return to them again in the initial sections of the corresponding Chaps. 4, 5 and 6.

2.1.1 Mesoscale and Synoptic Atmospheric Processes

Historically, one of the main classifying features of atmospheric processes is their horizontal spatial scale. So, e.g., according to the definition in (Khromov and Mamontova 1974), synoptic meteorology is a doctrine of atmospheric macroscale processes and weather prediction based on their study. Thus, synoptic processes are characterized by maximum (macroscopic) scales and are opposed in this regard to processes of "intermediate scales" (mesoscale) and microscale.

Initially, the separation of atmospheric processes into synoptic and mesoscale was caused by a purely phenomenological approach to their description. The atmospheric processes which characteristics can be restored due to synchronous observations at several (about a dozen) stations of the synoptic network (with an average distance between stations of about 100 km) were referred to synoptic (from Greek "observed simultaneously"). This gave a characteristic scale of the order of 1000 km. Mesoscale processes, respectively, were called those whose horizontal dimensions are less than 1000 km, but are too large for observations at individual stations (at individual points) to provide an adequate description of them. The restriction of line of sight from a fixed point at the surface of the Earth led to a second boundary scale of about 10 km. As a result, the range of characteristic horizontal sizes of mesoscale processes was taken equal to 10–1000 km.

It is interesting that further progress in synoptic and mesometeorology and the use of additional physical considerations did not lead to a radical revision of this simple classification, although numerous, often mutually exclusive, proposals for its expansion and complication were put forward, see Fig. 2.1. So, in (Orlansky 1975; Khromov and Petrosyants 2006), atmospheric processes with a characteristic horizontal scale of 200–2000 km and a lifetime of the order of days and weeks are classified as mesoscale, and in (Pielke 2002; Thunis and Bornstein 1996)—as synoptic (macroscale). Examples of atmospheric phenomena of this magnitude include tropical cyclones and atmospheric fronts (Veltishchev and Stepanenko 2006). Processes on a scale of 20–200 km are unanimously classified as mesoscale. A number of classifications include mesoscale processes with a characteristic horizontal size of 2–20 km (Orlansky 1975; Thunis and Bornstein 1996). All classifications specify scales over 2000 km as synoptic (macroscopic), while many distinguish proper synoptic (Khromov and Petrosyants 2006) or "macro-β" (Orlansky 1975; Thunis and Bornstein 1996) scale (up to 10,000 km) and global (Khromov and

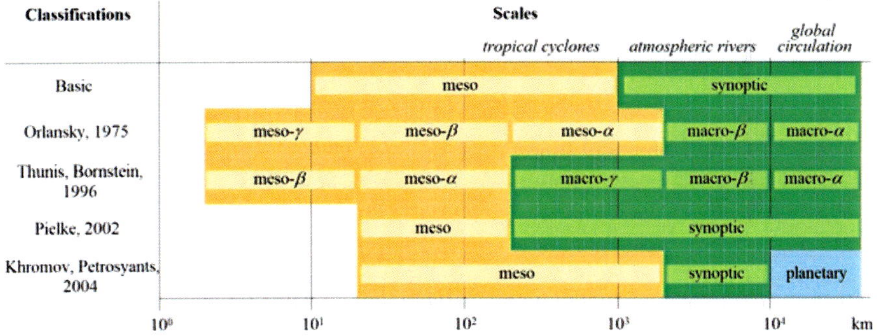

Fig. 2.1 Classifications of atmospheric processes by their spatial scales

Petrosyants 2006) or "macro-α" (Orlansky 1975; Thunis and Bornstein 1996) scale (over 10,000 km).

The scheme gives a simple and sufficient picture of multiscale atmospheric processes for the purpose of the book. For a deeper study of the issue, the reader can be addressed to (AMS Glossary 2020), as well as (Russian Hydrometeorological 2009; Golitsyn 2013).

The variety of atmospheric processes of different scales does not allow focusing on each of them. In this regard, the following approach was adopted in the book. A tropical cyclone was chosen as a prominent representative of mesoscale phenomena. Chap. 4 is devoted to the study of tropical cyclones based on satellite radiothermovision. Similarly, the so-called "atmospheric rivers" are considered as phenomena of a synoptic scale in Chap. 5. To demonstrate the wide range of satellite radiothermovision capabilities, Chap. 6, based on this approach, analyzes the characteristics of global atmospheric circulation over 17 years (2003–2019). Thus, in the framework of a unified research methodology, almost the entire range of actual horizontal scales of 10^2–10^4 km is covered. There are no fundamental methodological limitations for the transition to scales of the order of 10^1 km. This requires an increase in the detail of satellite radiothermal measurements in spatial resolution (up to units of kilometers) and periodicity of review (up to units of hours), which is in line with current global trends in the development of remote sensing of the Earth from space.

2.1.2 *Mesoscale: Tropical Cyclones*

Tropical cyclones are atmospheric phenomena of a catastrophic nature, regularly leading to human casualties and serious damage to coastal infrastructure (Pielke and Pielke 1997; Sharkov 2000, 2012; Emanuel 2003; Timofeev 2009). Along with the most famous distinguishing feature of mature tropical cyclones—a hurricane wind exceeding (sometimes substantially) 33 m/s, two other immediate danger factors are the collapse of a storm (surge) wave and heavy precipitation, causing a cascade of natural disasters when the cyclone breaks on land.

A characteristic feature of the structure of a tropical cyclone that distinguishes it from other tropical disturbances is the presence of a central warm core (Palmén and Newton 1969; Emanuel 2003). A mature tropical cyclone can be divided into the following four regions (Palmén and Newton 1969): "(1) an outer region with inward-increasing cyclonic increasing wind speed and limited convection; (2) a belt, in the inner portion of which the wind reaches hurricane strength, characterized by squall lines and heavy convection; (3) a more or less ring-formed inner rain region with very heavy rain and squalls and maximum wind strength; and (4) the eye inside a "transition zone" through which there is a rapid inward decrease of wind speed".

A brief chronology of scientific studies of tropical cyclones is given, for example, in (Emanuel 2003; Sharkov 2005). Due to their characteristic scales and formation features (above the open tropical ocean), science for a long time could not recreate a holistic picture of these peculiar and formidable natural phenomena. It is interesting to note that some historical artifacts can be interpreted as evidence of the familiarity of the ancient, "primitive" civilizations of America with the spiral structure of a tropical cyclone (lecture by K. Emanuel at Cornell University "How hurricanes respond to climate change" (Emanuel 2017).

Despite the dramatic progress in the study of tropical cyclones, provided by the development of satellite remote sensing technologies, a large number of open problems remain even in fundamental issues of the origin, evolution, formation of the trajectory of tropical cyclones (Emanuel 2003; Sharkov 2010). In particular, attempts to develop a "local" prognostic criterion for the so-called rapid intensification of a tropical cyclone (Kaplan et al. 2010), at which the maximum wind speed increases by 15 m/s or more per day, should be recognized as insufficiently satisfactory.

One of the fundamental aspects of the study of tropical cyclones is the study of energy exchange processes in the ocean-atmosphere system at spatial scales and time intervals of the order of a few hours characteristic of the structure of a tropical cyclone. It should be noted here that in most modern publications, close and, possibly, excessive attention is paid to the vertical flow of enthalpy, which characterizes the direct transfer of energy from the ocean to the atmosphere in an area limited by the diameter of the eye of the storm. Although this component is extremely important for the formation and maintenance of the general spiral structure of the cyclone, its contribution to the energy consumption and expenditure of a tropical cyclone is very insignificant (Palmén and Newton 1969), which will also be illustrated in Chap. 4 by specific examples. Moreover, already in early works on tropical cyclones, for example, (Ferrel 1856), it was indicated that the intensity and duration of a hurricane depend on the amount of water vapor entering the hurricane with lower air currents. The dominant role of advection of latent heat (horizontal inflow of water vapor) was also repeatedly emphasized in (Palmén and Newton 1969) and was demonstrated, inter alia, on the results of modeling of an "average tropical cyclone", as well as real hurricanes Desi and Helen. An unambiguous conclusion is that "...the essential source of total <tropical cyclone> energy is derived from the lateral influx of water vapor in the moist surface layer...", although "...the additional flux of latent and sensible heat from the sea un the core region represents a heat source not to be neglected". Thus, the spiral structure of a hurricane, under favorable conditions, provides compensation for dissipative energy losses (and even leads to its intensification) due to the convergence of latent heat, involving the surrounding air masses in the interaction, much more extended than the eye of the storm.

One of the probable reasons for insufficient attention to the study of latent heat advection processes from satellite data is the significant methodological difficulties of such an analysis. In the traditional, widely adopted "local" approach to the analysis of remote sensing data, measurements at various points are interpreted independently of each other. Generalization of the analysis results is performed at the final stage, as a rule, in the form of calculating the average values or other distribution characteristics of the reconstructed geophysical parameters. This approach is quite effective in solving a wide range of problems, including for estimating the instantaneous values of vertical flows of latent heat and enthalpy (Grankov and Milshin 2016). However, the study of latent heat advection requires restoration of the dynamic picture of the process, i.e. analysis of spatiotemporal relationships in the fields of geophysical parameters. The solution to this problem in the approach of satellite radiothermovision in relation to the study of the evolution of tropical cyclones is demonstrated in Chap. 4.

In conclusion of this section, it should also be noted that, in addition to the study of tropical cyclones as individual phenomena, considerable attention is paid to the study of global tropical cyclogenesis as a whole, as a complex multiscale process (Sharkov 2000; Emanuel 2003; Golitsyn 2008). Being an integral component of global circulation, tropical cyclogenesis, on the one hand, is influenced by general climatic changes (the manifestation of which can be expected, in particular, to change statistics on the frequency and intensity of emerging typhoons and hurricanes), and on the other hand, is actively involved in energy redistribution in the ocean-atmosphere system. In this regard, it is interesting, for example, the mentioning in a lecture by Emanuel (2017) that the average latitude at which tropical cyclones reach maximum intensity has recently shifted to the poles in both hemispheres (Kossin et al. 2014). Although the book briefly highlights the problems of a "global" approach to the study of tropical cyclogenesis, the main focus is on the study of individual tropical cyclones.

2.1.3 Synoptic Scale: Atmospheric Rivers

The term "atmospheric rivers" goes back to (Newell et al. 1992), in which the authors, noting the "filamentary structure" in the fields of diurnal streams of tropospheric water vapor, make comparisons with "rivers". The "rivers" ("tropospheric rivers") studied by the authors were extended (more than 2000 km in length) objects with a characteristic lifetime of several days, forming a moisture stream comparable in integral flow rate with the Amazon stream. Actually, this phenomenon was first called "atmospheric rivers", probably, in the subsequent work (Zhu and Newell 1994).

The authors investigated aspects of the interaction of atmospheric rivers (ARs) with extratropical cyclones and showed that a significant role in enhancing the latter is played by the release of latent heat, whose advection from tropical regions is provided by ARs. The research technique is noteworthy: based on the results of numerical modeling, the authors constructed vector maps of latent heat fluxes superimposed on color maps of atmospheric pressure for a fixed moment in time (noon and midnight at prime meridian); chronological sequences for individual basins of the World Ocean were collected from the obtained maps and analyzed as video streams.

Subsequently, numerical modeling revealed a significant role of AR as a factor of latent heat advection in the global atmospheric circulation system. So, in (Zhu and Newell 1998), it was concluded that ARs are responsible for "almost the entire" meridional transport of latent heat in mid-latitudes. From this point of view, further study of AR is of fundamental scientific interest.

Although the term "atmospheric rivers" still raises some objections, and a number of alternatives have been proposed by meteorologists (Bao et al. 2006; Knippertz and Wernli 2010), the phenomenon itself has attracted more and more attention in recent decades and has become the object of comprehensive research based on both modeling and analysis of remote data, as well as a combination of these approaches. In particular, it was shown that a significant part of extreme weather events (storm winds, torrential rains, snowfalls, floods, mudflows) in coastal continental zones of middle latitudes is associated with the effect of AR landfall (Ralph et al. 2006; Leung and Qian 2009; Guan et al. 2010; Ralph and Dettinger 2011). This indicates the great practical significance of the study of AR.

The mechanisms of the formation and evolution of ARs are the subject of active research, and so far, there remain problematic aspects of their realistic reproduction in numerical models (Gimeno et al. 2014). The task of the systematic study of ARs on climatically significant scales and the related problem of automatic detection of ARs and the diagnosis of their parameters against other atmospheric processes has become extremely urgent. Approaches to solving the latter problem are developing in two main directions (Gimeno et al. 2014): analysis of the spatial structure of the field of total precipitable water according to remote sensing data (Ralph et al. 2004; Matrosov 2013), modeling or reanalysis (Dettinger et al. 2011); and analysis of vertically integrated advective water vapor flows using numerical models (Zhu and Newell 1998). Moreover, as noted above, an integrated approach that combines the use of remote data and model estimates is becoming more common (Wick et al. 2013).

In Wick et al. (2013), the following main problems are identified that arise when using satellite radiometry data in the study of climatology of ARs:

(1) data gaps in some cases significantly complicate the automatic detection of AR;
(2) the numerical criteria for detecting ARs, developed on a limited scope of observations over individual basins of the World Ocean, require verification and refinement for universal use on a global scale;
(3) to improve the quality of detection of AR, it is desirable to have synchronous estimates of the fields of total precipitable water and advection of latent heat.

Chapter 5 of the book, in particular, shows how these problems find a solution in the framework of the satellite radiothermovision approach.

It should also be noted that until recently, researchers focused on the ARs over the northeast Pacific Ocean in connection with their impact on the weather and climate of the western United States. With a significant lag, the study of ARs over the North Atlantic began, although their impact on the weather and climate of Europe and the Arctic can be no less significant. The least studied are the ARs over the Indian Ocean (potentially capable of extending their influence to the Far Eastern coast of Russia) and the southern hemisphere. In this regard, universal analysis methods, which allow, within the framework of a unified approach, to study the ARs simultaneously over all the basins of the World Ocean, are becoming most relevant.

2.1.4 Planetary Scale: Global Atmospheric Circulation

The consideration of global atmospheric circulation as another object of research in the book has two objectives. Firstly, this allows demonstrating the approach and capabilities of satellite radiothermovision in relation to atmospheric processes of the largest scales of the order of 10^4 km. Secondly, in contrast to previous cases, where the small size and (or) the existence of the phenomena determined a certain selectivity of data processing, in the case of analyzing global circulation, the most complete analysis of the available information array of satellite radiothermal monitoring data was necessary. Such an analysis most clearly shows the universality of the applied approach. It is clear that it is fully applicable to the tasks of regional analysis, when processing data of higher spatiotemporal detail, to highlight individual small-scale factors against the background of large-scale synoptic processes.

An important advantage of global atmospheric circulation as an object of study is that its average characteristics, as well as their typical seasonal variations, are for a long time the subject of careful study on the basis of long-term, constantly expanding sets of various experimental data and increasingly detailed results of modeling (Palmén and Newton 1969; Peixoto and Oort 1983; Liu and Tang 2005; Robertson et al. 2014). In this sense, the calculations carried out within the framework of satellite radiothermovision, regardless of previous studies, have a reliable foundation for verification and confirmation, identifying the advantages and problem areas of the implemented analysis methodology, and detecting possible trends of various parameters of general circulation when comparing long time series.

On the other hand, it should be emphasized that changes in the structure and intensity of global atmospheric circulation are among the most important indicators of variations in the Earth's climate (Reichler 2009; Pan et al. 2017). An extremely actual area of research is, in particular, a comprehensive (based on generalization and analysis of remote data, reanalysis and theoretical-numerical modeling of climate) studies of the relatively recent trend of "widening of the tropics" (Reichler 2009). This tendency consists in shifting the boundaries of the Handley cells and the associated jet streams to the poles in both hemispheres and is clearly traced in the time series of various satellite remote sensing data (see Chap. 6), however, it is not yet well reproduced by modern climate models. Therefore, works that develop new approaches to a detailed analysis of the structure and evolution of the global atmospheric circulation based on remote data are extremely relevant.

It must be borne in mind that the process of "widening of the tropics" in turn is capable of causing further climatic changes that significantly affect the state of the Earth's global ecosystems. Changes in the meridional distribution of middle zonal winds, a shift of westerly winds to high latitudes will lead to desertification of the arid territories of the African and American continents (Hu and Fu 2007; Reichler 2009). Other ecosystems will also undergo changes, up to polar latitudes, in particular, due to a shift in the characteristic trajectories of extratropical cyclones, as well as deeper and more frequent penetration of tropical cyclones into high latitudes. The latter trend, as indicated in paragraph 2.1.2, is, in fact, already noted. Of course, the climate of the Arctic will also be affected by these changes, which is one of the key factors of not only the regional, but also the global state of the Earth's ecosystems. In this regard, some aspects of the study of the Arctic region by means of satellite radiothermovision are considered in the dissertation specifically, albeit in a brief form.

The general structure of the atmospheric circulation and its global dynamics are clearly manifested in the fields of total precipitable water, which can be reliably reconstructed from satellite radiometry data. In Chap. 6, the satellite radiothermovision approach is applied to reproduce a qualitative picture and a number of specific quantitative characteristics of the general atmospheric circulation based on an analysis of the processing products of seventeen-year continuous global satellite radiothermal observations.

2.1.5 The Commonality of Research Objects of Various Scales in the Context of Satellite Radiothermovision

It should be emphasized that all the objects of research considered in the book have common features from the point of view of the main physical mechanism—the advection of latent heat in the lower troposphere—which determines their formation (atmospheric rivers), evolution (tropical cyclones), or a number of the most important large-scale characteristics (global circulation). In the context of applied tasks, it is important to recall once again that tropical cyclones and atmospheric rivers pose a serious threat to the population and infrastructure of coastal areas. At the same time, due to the laws of general circulation, impacts of tropical cyclones are mainly affected by the eastern coasts of the continents, and, on the contrary, mainly the western coasts suffer from atmospheric rivers. One of the most important (both applied and fundamental) aspects of the connection of these processes with global circulation and climate variations is the transfer of atmospheric moisture and latent heat carried out by them. The role of atmospheric rivers in the meridional transport of tropospheric moisture was emphasized above. Determining the contribution of global tropical cyclogenesis to this process remains an open problem. It should be noted that the change in the frequency and intensity of extreme and catastrophic weather events is considered by most researchers as an important indicator associated with climate change and reflecting the processes of large-scale restructuring of the atmospheric circulation. A deep study of such relationships is possible based on an analysis of the energy balance in the ocean-atmosphere system, one of the most important components of which is latent heat.

Obtaining, in the required volume and with sufficient spatiotemporal resolution, adequate information on the state of the geophysical fields of the ocean-atmosphere system and, in particular, on the lower troposphere, is ensured in the work performed by the attraction and efficient use of the available data from satellite radiothermal remote sensing and their processing products. A brief overview of the methods developed in remote sensing for reconstructing the geophysical fields of the lower troposphere and the ocean-atmosphere boundary is given in the next section of this chapter. The third section of the chapter is devoted to the description of algorithms for calculating the dynamic characteristics of multidimensional physical fields, which are adapted in the basis of the satellite radiothermovision approach.

2.2 Passive Microwave Remote Sensing of the Atmosphere in Short

The effectiveness of the microwave radiometric method for studying the Earth's atmosphere (and, in particular, the lower troposphere) is due to the combination of high transparency of the atmosphere in wide ranges of the microwave spectrum and the lines of selective radiothermal radiation (and absorption) of some of its gas components, primarily oxygen and water vapor. The features of the spectrum of the atmosphere's own microwave radiation are determined both by the processes of radiation generation in all its layers and by the attenuation (absorption and scattering) of radiation along the entire path of its propagation. Thus, the measurement of the spectrum of the radiothermal radiation of the atmosphere makes it possible to extract quantitative information about the three-dimensional fields of its geophysical parameters (temperature, gas composition, hydrometeors).

Mathematically, the relationship between the radiobrightness spectrum and the distribution function of these parameters along the radiation propagation path is described by convolution of the form of the Fredholm integral equation, see, for example, (Ulaby et al. 1981; Sharkov 2003; Armand and Polyakov 2005; Kutuza et al. 2016). As a result, the task of reconstructing the three-dimensional structure of the atmosphere in the general case is mathematically incorrect: essentially different spatial distributions of atmospheric parameters that give indistinguishable, within the limits of measurement errors, radiobrightness spectra are hypothetically acceptable. Overcoming this problem with the help of additional information is possible on the basis of a mathematical apparatus developed in (Tikhonov and Arsenin 1977). Thus, the method of statistical regularization has been developed in relation to problems of remote sensing of the atmosphere in different spectral ranges (Kondratyev and Timofeev 1970; Basharinov et al. 1974; Troitsky et al. 1993; Rodgers 2000). An approach has been developed based on additional information on temporal variations in temperature and heat fluxes on the interface between the atmosphere and the underlying surface, "thermal history" (Gaikovich 1994, 1996, 1999, 2003) Other promising alternatives have been proposed, for example, a phenomenological approach using the apparatus of artificial neural networks (Blackwell 2005; Polyakov et al. 2014). Considerable attention is paid to improving the instrumental and methodological base of experimental research, for example (Kadygrov and Pick 1998; Westwater et al. 1999; Kadygrov et al. 2015). The problem continues to remain in the focus of relevant scientific and technical research (Sterlyadkin et al. 2017).

The situation is drastically simplified if integral parameters of the atmosphere are of interest, such as total precipitable water (total mass of water vapor in a vertical atmospheric column with a unit base area) and total liquid water content of clouds. In this case, on the basis of a general physical model for the propagation of microwave radiation, it is possible to construct simplified regression relationships between the sought parameters and the recorded radiobrightness temperatures of the atmosphere. Clarification ("tuning") of the proposed relationships is carried out by comparing the results of real radiothermal measurements with independent (for example, radiosonde) data. The extensive construction of all-weather algorithms, in particular, by soviet scientists was greatly stimulated by preparation and analysis of results of "Kosmos-243" and future space experiments, and was associated with the need to take into account the influence of dense cloudiness (Basharinov and Kutuza 1968; Akvilonova and Kutuza 1978; Kutuza and Smirnov 1980) and precipitation (Kutuza 1968; Smirnov 1984; Kutuza et al. 1998; Ilyushin and Kutuza 2016).

When observing from aerial or space-borne platforms, it is also necessary to take into account the contribution to the recorded signal of the own and scattered radiation from the underlying surface. The most reliable estimates of the integral parameters of the atmosphere are obtained over the ocean. One of the main reasons for this is the possibility of constructing a fairly accurate and universal low-parametric model of the influence of the ocean surface on the recorded radiobrightness spectrum of upward radiation (the key parameters are the speed of the surface wind, as well as the temperature and salinity of the surface layer of the ocean) (Hollinger 1971; Wu and Fung 1972; Pereslegin 1975; Sasaki et al. 1987; Trokhimovski et al. 1995, 2003; Johnson 2006). In addition, the own microwave radiation of a foam-free and ice-free water surface is, as a rule, substantially less (a characteristic emissivity of about 0.5) than of land (emissivity may exceed 0.9) at the same thermodynamic temperature. In the future, except for special cases, we will talk exclusively about radiothermal observations of the ocean-atmosphere system and fields of atmospheric parameters over the ocean.

A fundamental role in solving the problems posed in the book is played by a detailed analysis of the state and evolution of global and regional (over individual water areas) fields of the total precipitable water. In this connection, of central interest are the accuracy of reconstructing the total precipitable water fields from satellite radiothermal measurements, the detail and completeness of observations (characteristic dimensions of the spatial element, the frequency of the survey of a given area, the degree of coverage with measurements of all basins of the World Ocean per day), as well as their long-term stability—the ability to build long, uniform quality chronological data series. A description of the current state of art in this area should be preceded by a brief historical review.

2.2.1 Brief Historical Overview

As is known, the rapid progress of passive microwave remote sensing at the initial stage was facilitated by the experience of experimental and theoretical studies in the field of radioastronomy, where the influence of the atmosphere was considered as an interfering factor requiring accounting and compensation (Ulaby et al. 1981; Njoku 1982; Sharkov 2003; Armand and Polyakov 2005; Kutuza et al. 2016). The pioneering study of the atmosphere as such using microwave radiometry was the work of Dicke (1946), Dicke et al. (1946). The historic milestone was the first experience of radiothermal observations of the Earth from a satellite made in the USSR—"Kosmos-243", 1968, "Kosmos-384", 1970 (Basharinov et al. 1974; Gurvich and Kutuza 2010; Gorbunov and Kutuza 2018; Kutuza et al. 2019).

In (Basharinov et al. 1974, p. 168), the first experiments are described of constructing maps of the total precipitable water of the atmosphere Q over the oceans based on satellite track measurements. The reconstructed Q values at the nodal sub-satellite points, selected in 200 km increments along the projections of several consecutive orbits, formed the basis for constructing the field isolines. An additional correction was provided by data from radiosonde observations. Among other things, the dynamics of the Q fields was investigated by comparing maps related to several consecutive days of observation. In particular, it was concluded that, although such maps represent, in essence, "transects" of the observed total precipitable water field in time (measurements on adjacent passes are separated by an interval of 1.5 h), they can be processed and analyzed as a whole as a first approximation (ibid., p. 169). It should be noted the low spatial detail of the obtained maps, caused, inter alia, by the imperfection of the observational technique at a fixed (at nadir) viewing angle.

Already on the Nimbus-5 satellite, launched by the USA in 1972, an ESMR radiometer was installed, which carried out electronic scanning in a plane perpendicular to the flight path, ensuring a bandwidth of 3100 km. However, the ESMR had only one measuring channel at a frequency of 19.35 GHz, and its data were mainly used to obtain a qualitative picture of the spatial distribution of intense precipitation and sea ice (Petty 1990). Significantly more attention was drawn to the NEMS multichannel radiometer data (Staelin et al. 1975, 1976; Grody 1976), installed on the same Nimbus-5 and, in fact, reproducing the nadir observation scheme, first implemented on the Kosmos-243. A serious analysis was made of the possibilities of restoring the total precipitable water of the atmosphere, detecting dense clouds and areas of precipitation, determining the liquid water content of clouds, predicting the development of storm activity, and studying the radiation budget of the Earth (Chang and Wilheit 1979; Petty 1990). Broad prospects were revealed for the joint analysis of ESMR and NEMS data for a significantly more accurate restoration of the geophysical parameters of the atmosphere and the state of the sea surface (Staelin et al. 1976; Chang and Wilheit 1979), which ultimately led to the development of the concept of inclined (i.e., at viewing angles that deviate significantly from the nadir) two-polarization multi-frequency measurements.

An experiment carried out by the USSR in 1974–1976 on the Meteor-18 satellite became the next development step, which implemented microwave radiometric polarization measurements. According to measurements from Meteor-18 over the sea surface and ice fields, the total precipitable water and cloud liquid water content were estimated, precipitation detection, determination of the type and position of fronts, localization of precipitation zones and their distribution over the area were performed (Gorelik et al. 1975).

Launched in 1975 on the Nimbus-6 satellite, the new ESMR and SCAMS instruments embodied a number of ideas formulated by the results of the first experiments. The SCAMS scanning microwave spectrometer was a significant development of NEMS, which scanned perpendicular to the direction of flight in a swath of 2400 km at several frequencies, Fig. 2.2. The new ESMR for the first time implemented conical scanning geometry (with a fixed inclination of the line of sight to the local normal) in a swath 1270 km wide, performing bipolarization measurements at a frequency of 37 GHz. The joint use of SCAMS and ESMR measurement data made it possible, in particular, to significantly improve the accuracy of the estimates of the total precipitable water, as shown by comparisons with the radiosonde data (Grody et al. 1980).

In 1978, a new generation instrument was put into operation—the Scanning Multichannel Microwave Radiometer, SMMR (Gloersen and Barath 1977). Like the last ESMR, it carried out bipolarization measurements with a conical scanning geometry (for the first time scanning was realized mechanically, due to the rotation of the antenna). As in NEMS and SCAMS instruments, the measurements were multifrequency; the frequency set was optimized taking into account previous experience and new tasks, see Fig. 2.2. SMMR was designed to obtain estimates of ocean surface temperatures, surface wind speed, ice cover area and ice age, total precipitable water and total cloud liquid water content, precipitation intensity and some other geophysical parameters, primarily in the interests of meteorology, oceanographic research, and for weather forecasting. Another key innovative step was the simultaneous launch of two identical copies of the device on the satellites Seasat-A and Nimbus-7. It was the operating experience of this device, at least in the framework of the American program for the development of radiothermal satellite remote sensing, that allowed identifying and, over time, eliminating a number of conceptual shortcomings both in the instrument base and in approaches to solving inverse problems when restoring the geophysical parameters of the Earth's ocean-atmosphere system (Petty 1990).

A turning-point event is the inclusion in 1987 of scanning radiometers SSM/I (Special Sensor Microwave/Imager) in the satellite equipment of the long-term

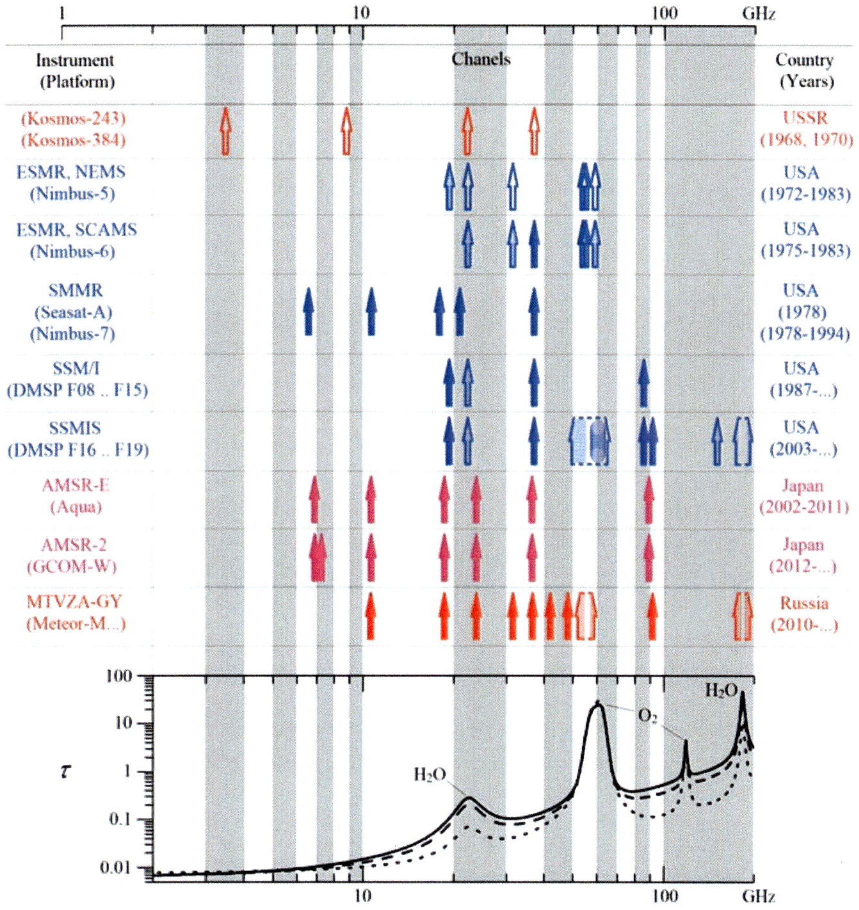

Fig. 2.2 Location of the measuring channels of some satellite microwave radiometers (arrows) relative to the spectrum of the optical thickness of the atmosphere in the range 1–200 GHz: empty arrows—measurements at nadir; with vertical/horizontal/circular shading—measurements on vertical/horizontal/circular polarizations during transverse scanning; shaded arrows—bipolarization measurements during transverse scanning. At the bottom of the graph is the spectral variation of the characteristic optical thickness of the standard atmosphere τ: the solid line is for tropical latitudes; long dotted line—for temperate latitudes in summer; short dash—for temperate latitudes in winter

research program DMSP (Defense Meteorological Satellite Program of the USA). The main task to be solved with their help was the restoration of global distributions of the integral parameters of the atmosphere over the ocean (total precipitable water, cloud liquid water, precipitation intensity), as well as scalar fields of the surface wind speed. At the same time, SSM/T scanners were installed on the same satellites, aimed at restoring the vertical temperature profiles of the atmosphere.

Subsequently, in accordance with the program, the functionality of SSM/I and SSM/T was combined in the SSMIS scanning and sounding radiometers (since 2003). SSM/I was the first serial satellite radiometer to provide continuous monitoring of the Earth for several decades. Since 1991, at least two (since 1995, three) copies of SSM/I (SSMIS) instruments have been continuously operating in orbit simultaneously. This provided access to a qualitatively new level of radiothermal satellite monitoring of the Earth, including in aspects of calibration of measurements and restoration of geophysical parameters of the ocean-atmosphere system (Hollinger 1988; Petty 1990).

The progress in the accuracy of reconstructing the integral geophysical parameters of the atmosphere was naturally associated with the development of the instrument base and an increase in the volume of satellite measurements available for analysis.

So, in (Basharinov et al. 1974) and a number of earlier works described in this monograph, linear regression was used that relates the total precipitable water of the atmosphere in g/cm^2 and the brightness temperature of the outgoing radiation measured at nadir at a wavelength of 1.35 cm. Regression coefficients were estimated by comparing model calculations with radiosonde data. In particular, to analyze the results of satellite measurements in (ibid.), the following relation was used:

$$T_{1.35} = 18Q + 127. \tag{2.1}$$

The possibility of correcting estimates (2.1) in the presence of cloudiness using measurement data on a wave of 0.8 cm, was also considered there. As a result of the application of the algorithm for joint processing of measurements, the standard deviation of the estimates from satellite data from the results of radiosonde measurements (in a sample of 38 sessions in the absence of precipitation) was 0.2 g/cm^2 (2 kg/m^2) (ibid., p. 164).

Similar regressions were also used in the analysis of radiothermal measurements from the first American satellites Nimbus-5 and Nimbus-6. However, several alternative algorithms had already been proposed for the interpretation of SMMR data, using a regression relationship between the integrated geophysical parameters of the atmosphere and nonlinear functions of brightness temperatures (Wilheit and Chang 1980; Petty 1990). One of the first attempts to introduce non-linearity by simultaneously (empirically) taking into account regional and climatic features was carried out in the so-called Hughes algorithms (Petty 1990), proposed for the analysis of SSM/I measurement data before its launch. These algorithms modeled nonlinearity using piecewise linear regressions, the coefficients of which were functions of geographical coordinates. Soon after receiving the first SSM/I data, it was found that Hughes algorithms could not provide the initially stated accuracy of the estimates of the integral geophysical parameters of the atmosphere (ibid.). As a result, three main groups of approaches to the introduction of nonlinearity were formed. Two of them differ in the type of functional connection: "logarithmic" approaches construct regressions between the desired geophysical parameters and the logarithms of brightness temperatures (Wilheit and Chang 1980; Schluessel and Emery 1990; Petty 1990, 1994; Lojou et al. 1994; Wentz 1995; Ruprecht 1996), "polynomial" use power polynomials from brightness temperatures (Alishouse et al. 1990; Sun and Weng 2008). Combinations of these approaches are also used, for example (Schluessel and Emery 1990). Below, as an example, some of the proposed regression expressions for the total precipitable water (kg/m^2) are listed, obtained in closed form relative to the brightness temperatures in the SSM/I channels: T_{19V}, T_{19H}, T_{22V}, T_{37V}, T_{37H}, T_{86V} (K), where the subscript indicates the (rounded) frequency in GHz and polarization of received radiation.

So, according to (Petty 1990),

$$Q = 11.98\ln(280 - T_{19V}) + 42.06\ln(280 - T_{19H}) - 54.36\ln(280 - T_{22V}) - 20.5, \quad (2.2)$$

while in (Petty 1994)

$$Q = 171.4 + 4.638\ln(300 - T_{19V}) - 61.76\ln(300 - T_{22V}) + 19.58\ln(300 - T_{37H}). \quad (2.3)$$

In (Ruprecht 1996), the calculation is as follows:

$$Q = 131.95 - 39.50\ln(280 - T_{22V}) + 12.49\ln(280 - T_{37V}), \quad (2.4)$$

and in (Wentz 1995) [cited in (Sohn and Smith 2003)] it is proposed:

$$Q = 88.76 + 43.289\ln(290 - T_{19V}) - 62.217\ln(290 - T_{22V}), \quad (2.5)$$

In (Schluessel and Emery 1990) the following regression relations are given

$$Q = 22.15 - 4.116\ln(280 - T_{22V}) - 0.001777(T_{19V} - \ln(280 - T_{22V})), \quad (2.6)$$

$$Q = 23.82 - 4.059\ln(280 - T_{22V}) + 0.02451(\ln(280 - T_{22V}) - T_{37V}), \quad (2.7)$$

$$Q = 22.73 - 3.969\ln(280 - T_{22V}) - 0.03423\left(T_{37V} - T_{86V}^{0.8}\right). \quad (2.8)$$

In (Lojou et al. 1994) a regression is constructed

$$Q = 20.75 - 2.582\ln(280 - T_{19H}) - 0.3919\ln(280 - T_{19V}) - 3.610\ln(280 - T_{22V}) + 2.729\ln(280 - T_{37H}) - 0.5118\ln(280 - T_{37V}) \quad (2.9)$$

Finally, in (Alishouse et al. 1990) the following expression is given

$$Q = 232.894 - 0.149T_{19V} - 1.829T_{22V} - 0.370T_{37V} + 0.0062T_{22V}^2, \quad (2.10)$$

and in (Sun and Weng 2008) it was additionally corrected in order to increase the accuracy of estimates in extreme weather conditions in the form

$$Q = -3.753 + 1.507Q_A - 0.1933Q_A^2 + 0.00219Q_A^3, \tag{2.11}$$

where Q_A is the value obtained in the calculation according to Eq. (2.10).

It should be noted that the publicly available version of the original paper (Alishouse et al. 1990) made a typo in the sign of the last regression coefficient. Corrected equations can be found, for example, in (Sohn and Smith 2003; Sun and Weng 2008).

The third group of approaches implements the method of successive approximations of the values of the desired geophysical parameters, which give the best correspondence between the actually measured and simulated values of radiobrightness temperatures, for example (Wentz 1997), see also (Petty 1990; Sohn and Smith 2003). In this case, one cannot give an explicit expression for Q, since there is only a model of the radiobrightness spectrum of the ocean-atmosphere system parameterized by the desired geophysical parameters, and an algorithm for reversing the system of equations to reconstruct the values of these parameters at given values of the measured radiobrightness temperatures.

The diversity of the described approaches necessitated a thorough mutual comparison both to select the most optimal ones, to identify the causes of mutual discrepancies, and, if possible, eliminate existing errors and obtain more accurate solutions (Wentz 1995; Hollinger 1988; Sohn and Smith 2003; Sun and Weng 2008). To date, one of the most widely recognized as a reliable global (that is, universally applicable in all areas of the ocean-atmosphere system) algorithm is the one developed in (Wentz 1997). It is regularly tuned and updated, including for adaptation to the data of new satellite radiometers. In fact, it is a documented standard adopted for streaming processing of satellite radiothermal Earth monitoring data at RSS (Remote Sensing Systems, USA). Updated versions of the algorithm are used not only for processing newly received information, but also for reprocessing archived data, which ensures the construction of continuous long-term series of global observations of integral geophysical parameters of the atmosphere, including in the interests of climate research.

2.2.2 Present State

The DMSP program remains a unique space mission that provides global, regular, long-term, uniform informational quality of satellite radiothermal monitoring of the Earth. Based on measurements of the SSM/I and SSMIS radiothermal fields of the ocean-atmosphere system, a number of specialized electronic archives have been formed and are being maintained, both abroad and in Russia. RSS is implementing a long-term project for maintaining and promptly replenishing the electronic archive of satellite-based radiothermal monitoring data processing products. This archive contains, in particular, global fields of total precipitable water, cloud liquid water content and other geophysical parameters calculated on a regular geographical grid with a step of 0.25°. The archive information covers the interval of continuous radiothermal satellite observations since July 1987 and is currently open for use for scientific purposes without any restrictions. The quality of information has been widely recognized by the world scientific community, and RSS products are in demand in solving a wide range of meteorology and climatology problems.

Of course, focusing on data from a unique space mission is a source of vulnerability for any long-term research program. So, in February 2016, the DMSP F19 satellite (launched last in the series in 2014) was lost due to problems with the control and monitoring unit. The launch of the next satellite, previously scheduled for 2020, probably will not take place until 2023 due to funding problems (DMSP F19 2016). Still, quite comprehensive operational information is obtained thanks to the previously launched satellites F16, F17, F18, which operating time in orbit has already significantly exceeded the five-year warranty period.

In this situation, the efforts of other countries aimed at developing their own or joint missions of space radiothermal monitoring of the Earth become especially significant. In the context of the topic of the book, space research programs of Japan and Russia should be emphasized.

The AMSR series of scanning radiometers offers significant advantages over a number of technical characteristics over the SSM/I and SSMIS series. So, the AMSR-2 instrument, created by the Japanese Aerospace Agency (JAXA) in addition to measuring channels close to the SSM/I channels, had a number of low-frequency channels (6.925, 7.3, 10.65 GHz, see Fig. 2.2), which allowed obtaining more accurate information about the state of the underlying surface. Bipolarization measurements at all frequencies opened up additional possibilities for restoring the total precipitable water over land (Du et al. 2017). The large diameter of the antenna system's mirror (2 m) provided twice the spatial resolution than SSMIS (up to 5 km in the high-frequency channel 89.0 GHz) (Kutuza et al. 2016).

The AMSR-2 instruments are functioning in orbit since 2012 onboard GCOM-W satellites (currently, in May 2020, the GCOM-W2 satellite is operational). The launch of the next similar device on the GCOM-W3 satellite was also planned in 2020 according to (Observing Systems 2011). Its predecessors were the AMSR instrument, which had worked for a little less than a year (2002–2003) on the ADEOS-2 satellite, and the much more successful AMSR-E instrument created on the basis of cooperation with NASA and launched on the Aqua satellite. The term of the latter was more than 9 years (2002–2011).

Of significant interest is also a series of Russian-designed MTVZA scanners/probers (Cherny et al. 2003; Boldyrev et al. 2008; Cherniavsky et al. 2018). So, MTVZA-GY has 49 working channels in the range from 10 to 184 GHz. Currently (May 2020), one instance of these devices is operating on the satellite "Meteor-M" No. 2–2. The promising MTVZA-GY-MP device, the launch of which is planned after 2021, represents a new important stage in the development of this series. One of the problematic aspects of this space program is the lack of, at the moment, a developed infrastructure for user information support in interactive mode, which would provide quick access to the necessary data samples or the whole of them. In particular, this concerns the possibility of obtaining in the streaming mode products of deep processing of primary information brought to the level of fields of restored geophysical parameters. Providing this opportunity will make the described space program a serious alternative mission of DMSP in terms of satellite radiothermal monitoring of the Earth.

Another interesting Russian project is devoted to design of a microwave radiometer installable on the International Space Station (Sharkov et al. 2019).

More detailed reviews of historic (currently completed), modern, and prospective programs of satellite radiothermal sensing of the Earth can be found, for example, in (Kramer 2002; Kutuza et al. 2016).

The radically increased, in comparison with earlier experiments, volumes and quality of satellite radiothermal monitoring information, its successful and increasingly active application in the tasks of meteorology and climatology, made it possible to reach a new level of generalization of accumulated experience (Timofeev 2009; Observing Systems… 2011). Thus, the World Meteorological Organization (WMO) launched the OSCAR project (Observing Systems Capacity Analysis and Review Tool) (Observing Systems… 2011), which allows comparing the quality and volume characteristics of various Earth remote monitoring data and their processing products with the real needs formulated within the framework of various practical and fundamental tasks. Thus, all standard products of remote data processing, as well as the requirements for them, are "certified", which is, of course, an extremely important element of interdisciplinary cooperation. As an example, Table 2.1 below is compiled according to the WMO OSCAR "prerequisites" for total precipitable water.

The three values shown in each cell of Table 2.1 (except for the Tasks and Geographic Areas columns) define three levels of requirements (top to bottom

Table 2.1 Requirements for the parameters of satellite measurements of the atmospheric total precipitable water according to WMO OSCAR

Application	Accuracy (kg/m^2)	Horizontal resolution (km)	Periodicity	Timeliness	Coverage
Climate analysis	1.0 1.4 3.0	50 100 200	3 h 4 h 6 h	7 days 14 days 60 days	Global
Global NWP*	1.0 2.0 5.0	15 50 25	60 min 6 h 12 h	6 min 30 min 6 h	Global
High resolution NWP*	1.0 2.0 5.0	0.5 5 20	15 min 60 min 6 h	15 min 30 min 2 h	Global
Nowcasting	1.0 2.0 5.0	5 10 50	15 min 30 min 60 min	5 min 10 min 30 min	Global

* Numerical weather prediction

according to WMO classification): goal, breakthrough, and threshold. As the analysis of Table 2.1 shows, in terms of the accuracy of restoration of the total precipitable water, the methods of satellite radiothermal monitoring of the atmosphere reach the "goal" level, and in terms of spatial resolution they basically correspond or are close to the "breakthrough" one. The most urgent task is to ensure a higher periodicity of updating data on the state of the atmosphere (by increasing the constellation of remote sensing satellites, or—which is the subject of the book—by special tools for the dynamic analysis of available remote data). An important aspect is the timeliness of satellite information, i.e. its operational availability, especially in the tasks of short-term and ultra-short-term weather forecasting. Here, a significant role is played not only by the efficiency of the algorithms for processing primary information, but also by the developed infrastructure of information support for potential users.

It should be noted that the brief review presented concerned only the so-called global, i.e. universally applicable in different geographical areas and observation conditions, algorithms. In the tasks of regional analysis, there are wide possibilities for further improving the accuracy of the restoration of geophysical fields of the atmosphere by attracting additional information. As such, we can use both data from independent observations of the current state of the atmosphere and statistics of the characteristic distributions of the parameters of the ocean-atmosphere system and (or) correlation between them in a given region (Mitnik and Mitnik 2003). Similar approaches are applicable for more accurate diagnostics of a narrow class of states of the ocean-atmosphere system, for example, under extreme weather conditions (Zabolotskikh et al. 2013). However, as indicated above, the accuracy of global algorithms seems to be quite sufficient when solving the problems discussed in the book. A dynamic analysis of geophysical fields was of central interest, and the problems associated with it were discussed in the next section and in Chap. 3. However, there are no methodological limitations for using the results of combined calculations using various algorithms and data from different instruments in the framework of satellite radiothermovision, which was demonstrated in the book using a multisensory approach.

2.3 Instruments and Data to Our Service

Earth's satellite radiothermal remote sensing is developing at an increasing rate. The pioneering solutions, which provided the first radiometric observations from satellites to nadir, first evolved into multichannel *imagers*, which allowed reconstructing two-dimensional (height-integrated) geophysical fields of the atmosphere, and then into *imagers-sounders*, which opened up prospects for obtaining a complete three-dimensional picture of the state of the atmosphere. Active work is underway on passive aperture synthesis methods aimed at a significant improvement in the spatial detail of the information received.

Along with new developments, the serial launch of reliable, well-proven instruments is important. It provides long-term homogeneous series of observations of the Earth of high quality and a smooth transition to the use of information from next-generation devices due to the long time intervals of their joint work. This opens up wide prospects not only for weather, but also for climate studies, which are increasingly attracting various satellite data to the analysis.

For many reasons the available radiometric information is far from being fully used within the framework of the satellite radiothermovision approach. That is why it is even more important to take a look at the wide palette of modern capabilities, illustrating the development potential in numerous actual directions of satellite radiothermal monitoring of the Earth.

For the problems discussed in the book the spatial completeness of the Earth's coverage with microwave measurements is most important provided that retrieval of the corresponding geophysical parameters (first of all, the total precipitable water) is achievable. That is why the data of satellite microwave scanners are of primary interest. Next in priority are the data of microwave sounders, which provide information on the vertical structure of the atmosphere. It should be noted that many modern instruments to some extent combine the functions of a scanner and a sounder.

The WMO OSCAR system (Observing Systems… 2011) contains information on more than thirty types of satellite multi-channel microwave scanners and sounders (some of them were represented by several instruments operating in orbit simultaneously). Some of them combine the properties of a scanner and a sounder, while others have radiometric channels that are characteristic of only one type of device.

Of main interest for the development of satellite radiothermovision is information covering the widest latitudinal zone—up to the polar latitudes. Reasonable spatial resolution requirements are on the order of several tens of kilometers. The set of channels and the radiometric quality of the data should ensure the retrieval of the basic geophysical parameters of the atmosphere, first of all, the total precipitable water. In addition, for reasons of algorithmic simplicity, it is highly desirable to have the largest possible volume of measurements from satellites with close circulation periods (for example, from solar-synchronous orbits). Given these limitations, we obtain the following short list of devices that retained functionality by 2000 or later (according to available data as of May 2020), Table 2.2.

The first column of Table 2.2 lists satellite microwave imagers/sounders, while the other columns correspond to years starting from 2000. The color cells mark the years of operation of the listed devices: green color indicates devices with a set of frequency channels typical for imagers, yellow—for sounder, and blue for devices combining these two functions. The numbers in the cells give the maximum number of identical devices (if more than one) operating simultaneously this year.

The instruments differ in a set of radiometric channels, observation geometry, spatial resolution, geographical and daily coverage. Some of them for a long time functioned in orbit in several instances at once, which increases the completeness of observations. For more detailed information about the characteristics and operating time of satellite radiometric imagers and sounders, the reader can be addressed to the WMO OSCAR resource and the links contained therein.

The level of preliminary processing of the routinely accessible information of these instruments and the reliability of retrievals the geophysical parameters of the

Table 2.2 Some satellite microwave imagers/sounders operating in 2000–2020

Instrument	Year 20…																				
	00	01	02	03	04	05	06	07	08	09	10	11	12	13	14	15	16	17	18	19	20
AMSR																					
AMSR-2																					
AMSR-E																					
AMSU-A	1	2	2	2	1	2	2	3	3	4	4	4	4	5	5	4	4	4	4	5	5
AMSU-B		2	3	3	3	3	3	3	3	3	3	3	2	2							
ATMS																			2	2	2
Delta-2D																					
GMI (core)																					
HSB																					
MHS								2	2	3	3	3	3	4	4	4	4	3	3	4	4
MSMR																					
MTVZA																					
MTVZA-GY																					2
MTVZA-OK (MW)																					
MWHS-1												2	2	2	2	2					
MWHS-2																			2	2	2
MWRI														2	2	2	2	2	3	3	3
MWRI (HY-2)																					
SSM/I	6	5	5	5	5	5	5	4	4	3	3	3	3	3	3	3					
SSM/T-2	4	3	3	3	3	3	3	3	3	2	2	2	2	2	2	2					
SSMIS								2	2	2	3	3	3	3	4	4	4	3	3	3	2
WindSat																					

atmosphere over different types of underlying surface are also different. As a result, the data of some of them are priority in demand in satellite radiothermovision algorithms, the data of others are optionally used. Some of the available information currently does not find application in the framework of the approach considered in the book. Of course, one of the challenges is the further improvement of the methodology for including the full amount of data in the analysis according to the scheme described in the following chapters.

2.4 The Idea of "Dynamic" Data Analysis

Let's turn again to Table 2.1, which lists the basic requirements for the retrieved fields of total precipitable water. Correlation of these requirements with the modern capabilities of satellite radiometry shows that in a number of parameters (accuracy of estimates, spatial resolution), the necessary indicators are practically achieved. This underlines the prospects for further development of satellite microwave remote sensing.

The most critical issue is time resolution. So, a satellite in a sun-synchronous orbit provides Earth observation only twice a day—on the ascending and descending parts of the orbit. At the same time, gaps ("lacunae") remain between the swaths at low latitudes, which often cause significant difficulties in the analysis of mesoscale and synoptic atmospheric processes. The requirements for the time resolution indicated in Table 2.1 are units of hours or even tens of minutes with full global coverage of the Earth by observations. To achieve these aims exclusively with instrumental means, it is required to increase the number of devices simultaneously operating in orbit by an order of magnitude, and the quality of the final data set in terms of radiometric accuracy, stability, spatial resolution, and other parameters will be determined by the worst of these devices.

A practical alternative is to use the information redundancy of the data of existing devices. Indeed, at the scales under consideration, the fields of the observed geophysical parameters are characterized by spatiotemporal smoothness. This is already evident from a comparison of series of data obtained in one day from several satellites. Thus, the idea of introducing some simple kinematic model that describes the transformation of these fields in time arises. Calculation of the parameters of this model from observational data opens up new perspectives in the interpretation of satellite information. Knowing the parameters of such a kinematic model and using the data of real observations as “reference”, one can then interpolate (extrapolate) the geophysical fields to other points in time and in space. This, in principle, makes it possible to fill in data gaps (“lacunae”) and reconstruct the state of geophysical fields at times when real observations were not performed. In other words, it becomes possible to get as close as possible to the required indicators for temporal resolution and global spatial coverage, relying solely on the available satellite observation data. Moreover, once the dynamics of the (in terms of parameters of the kinematic model) is reconstructed the new opportunities appear: first of all, the opportunity to estimate vertically integrated fluxes of water vapor and latent heat at any point as well as through an arbitrary set boundary. Of course, this again raises the question of the accuracy of the estimates obtained.

The next chapter of the book is devoted to the development, justification, and algorithmic implementation of this idea within the framework of the satellite radiothermovision approach.

2.5 Concluding Remarks to this Chapter

1. Advection of latent heat in the lower troposphere of the Earth is an important factor determining the formation and (or) evolution of a wide class of mesoscale and synoptic atmospheric processes.
2. A practically non-alternative means of reliable and regular global diagnostics (monitoring) of the state and dynamics of the lower troposphere with spatial detail of the order of ten kilometers and a periodicity of at least twice a day are satellite radiothermal observations.
3. Of high relevance is the task of developing an approach to retrieving the characteristics of the dynamics of the lower troposphere, using satellite radiothermal remote data from solar-synchronous and other low orbits as input.

References

Akvilonova AB, Kutuza BG (1978) Radioteplovoe izluchenie oblakov (Microwave radiothermal emission of clouds). Radiotekhnika i elektronika 23(9):1792–1806 (in Russian)

Alishouse JC, Snyder SA, Vongsathorn J, Ferraro RR (1990) Determination of oceanic total precipitable water from SSM/I. IEEE Trans Geosci Remote Sens 28(5):811–816

AMS Glossary (2020) AMS Glossary of Meteorology (Electronic resource). http://glossary.ametsoc.org/

Armand NA, Polyakov VM (2005) Radio propagation and remote sensing of the environment. CRC Press, Boca Raton, 448p

Bao JW, Michelson SA, Neiman PJ, Ralph FM, Wilczak JM (2006) Interpretation of enhanced integrated water vapor bands associated with extratropical cyclones: their formation and connection to tropical moisture. Mon Weather Rev 134(4):1063–1080

Basharinov AE, Gurvich AS, Egorov ST (1974) Radioizlucheniye Zemli kak planet (Radioemission of the Earth as a planet). Nauka, Moscow, 187p (in Russian)

Basharinov AE, Kutuza BG (1968) Issledovanie radioizlucheniya i pogloshcheniya oblachnoi atmosfery v millimetrovom i santimetrovom diapazonakh voln (Investigation of radio emission and absorption of a cloudy atmosphere in millimeter and centimeter wavelengths). Trudy GGO 222:100–110 (in Russian)

Blackwell WJ (2005) A neural-network technique for the retrieval of atmospheric temperature and moisture profiles from high spectral resolution sounding data. IEEE Trans Geosci Remote Sens 43(11):2535–2546

Boldyrev VV, Il'gasov PA, Pantsov VY, Prokhorov YN, Strel'nikov NI, Cherny IV, Chernyavskii GM, Yakovlev VV (2008) Sputnikovyi mikrovolnovyi skaner/zondirovshchik MTVZA-GYa (Satellite microwave scanner/sounder MTVZA-GY). Sovrem Probl Distantsionnogo Zondirovaniya Zemli Iz KosmA 5(1):243–248 (in Russian)

Chang ATC, Wilheit TT (1979) Remote sensing of atmospheric water vapor, liquid water, and wind speed at the ocean surface by passive microwave techniques from the Nimbus 5 satellite. Radio Sci 14(5):793–802

Cherniavsky GM, Mitnik LM, Kuleshov VP, Mitnik ML, Cherny IV (2018) Mikrovolnovoye zondirovaniye okeana, atmosfery i zemnyh pokrovov po dannym sputnika "Meteor-M" N 2 (Microwave sensing of the ocean, atmosphere and land surface from Meteor-M No. 2 data). Sovremennye Problemy Distantsionnogo Zondirovaniya Zemli iz Kosmosa 15(4):78–100 (in Russian)

Cherny IV, Chernyavskii GM, Uspensky AB, Pegasov VM (2003) SVCh-radiometr MTVZA sputnika "Meteor-3 M" N 1: predvaritelnye rezultaty letnykh ispytaniy (Microwave radiometer MTVZA-GY on board Meteor-3 M satellite: preliminary results of flight tests). Issledovanie Zemli iz Kosmosa 6:1–15 (in Russian)

Dettinger MD, Ralph FM, Das T, Neiman PG, Cayan DR (2011) Atmospheric rivers, floods and water resources of California. Water 3(2):445–478

Dicke R (1946) The measurement of thermal radiation at microwave frequencies. Rev Sci Instruments 17(7):268–275

Dicke RH, Beringer R, Kyhl RL, Vane AB (1946) Atmospheric absorption measurements with a microwave radiometer. Phys Rev 70(5–6):340–348

DMSP F19 (2016) DMSP F19 military weather satellite declared lost after communications failure (Electronic resource). URL: http://spaceflight101.com/re-entry/dmsp-f19-military-weather-satellite-declared-lost-after-communications-failure/

Du J, Kimball JS, Jones LA, Kim Y, Glassy J, Watts JD (2017) A global satellite environmental data record derived from AMSR-E and AMSR2 microwave Earth observations. Earth Syst Sci Data 9(2):791–808

Emanuel K (2003) Tropical cyclones. Annu Rev Earth Planet Sci 31:75–104

Emanuel K (2017) How do hurricanes respond to climate change. Cornell University. URL: https://www.youtube.com/watch?v=lklAlBAOfkA

Ferrel W (1856) An essay on the winds and the currents of the ocean. Cameron & Fall, Book and Job Printers, 43p

Gaikovich KP (1994) Simultaneous solution of emission transfer and thermal conductivity equations in the problems of atmosphere and subsurface radiothermometry. IEEE Trans Geosci Remote Sens 32(4):885–889

Gaikovich KP (1996) Stochastic theory of temperature distribution and thermal emission of half-space with random time-dependent surface temperature. IEEE Trans Geosci Remote Sens 34(2):582–587

Gaikovich KP (1999) Spectral approach to correlation theory of thermal regime and thermal emission of the medium with random boundary conditions. IEEE Trans Geosci Remote Sens 37(1):3–7

Gaikovich KP (2003) Application of the correlation theory of thermal regime and thermal radio emission for atmosphere. IEEE Trans Geosci Remote Sens 41(12):2754–2759

Gimeno L, Nieto R, Vazquez M, Lavers DA (2014) Atmospheric rivers: a mini-review. Front Earth Sci 2. https://doi.org/10.3389/feart.2014.00002

Gloersen P, Barath FT (1977) A scanning multichannel microwave radiometer for Nimbus-6 and Seasat-A. IEEE J Oceanic Eng 2(2):172–178

Golitsyn GS (2008) Polar lows and tropical hurricanes: their energy and sizes and a quantitative criterion for their generation. Izv Atmos Ocean Physics 44(5):537–547. https://doi.org/10.1134/S0001433808050010

Golitsyn GS (2013) Statistika i dinamika pridodnyh protsessov i yavleniy: Metody, instrumentariy, resultaty (Statistics and dynamics of natural processes and phenomena: Methods, tools, results). KRASAND, Moscow, 398p (in Russian)

Gorbunov ME, Kutuza BG (2018) Cosmos-243 as the starting point for the development of microwave radiometry methods of the Earth's atmosphere and surface. Izv Atmos Ocean Physics 54(3):275–281

Gorelik AG, Dombkovskaya EP, Ozerkina VV, Semiletov VI, Skuratova IS, Frolov AV (1975) Mikrovolnovye polyarizatsionnye izmereniya na sputnike «Meteor» (Microwave polarization measurements on the "Meteor" satellite). Meteorol I Gidrol 7:36–45 (in Russian)

Grankov AG, Milshin AA (2016) Microwave radiation of the ocean-atmosphere: boundary heat and dynamic interaction, 2nd edn. Springer, Berlin,193p

Grody NC (1976) Remote sensing of atmospheric water content from satellite using microwave radiometry. IEEE Trans Antennas Propag 24(2):155–162

Grody NC, Gruber A, Shen WC (1980) Atmospheric water content over the tropical Pacific derived from the Nimbus 6 Scanning microwave spectrometer. J Appl Meteorol 19(8):986–996

Guan B, Molotch NP, Waliser DE, Fetzer EJ, Neiman PJ (2010) Extreme snowfall events linked to atmospheric rivers and surface air temperature via satellite measurements. Geophys Res Lett 37(20):L20401. https://doi.org/10.1029/2010GL044696

Gurvich AS, Kutuza BG (2010) "Kosmos-243" – pervyi v mire eksperiment po issledovaniyu Zemli iz kosmosa radiofizicheskimi metodami ("Kosmos-243"—the first experiment in the World on the Earth research from space using radiophysical methods). Issledovanie Zemli iz kosmosa 2:14–25 (in Russian)

Hollinger JP (1971) Passive microwave measurements of sea surface roughness. IEEE Trans Geosci Electron GE-9(3):165–169

Hollinger JP (1988) DMSP special sensor microwave/imager calibration/validation. Final report volume 1. Space sensing branch. Naval Research Laboratory, Washington, DC, 20375–5000, 190p

Hu Y, Fu Q (2007) Observed poleward expansion of the Hadley circulation since 1979. Atmos Chem Phys 7(19):5229–5236

Ilyushin YA, Kutuza BG (2016) Influence of a spatial structure of precipitates on polarization characteristics of the outgoing microwave radiation of the atmosphere. Izv Atmos Ocean Phys 52(1):74–81

Johnson JT (2006) An efficient two-scale model for the computation of thermal emission and atmospheric reflection from the sea surface. IEEE Trans Geosci Remote Sens 44(3):560–568

Kadygrov EN, Ganshin EV, Miller EA, Tochilkina TA (2015) Ground-based microwave temperature profilers: potential and experimental data. Atmospheric and Oceanic Optics 28(6):598–605

Kadygrov EN, Pick DR (1998) The potential for temperature retrieval from an angular- scanning single-channel microwave radiometer and some comparisons with in situ observations. Meteorol Appl 5(4):393–404

Kaplan J, DeMaria M, Knaff JA (2010) A revised tropical cyclone rapid intensification index for the Atlantic and Eastern North Pacific basins. Weather Forecasting 25(1):220–241

Khromov SP, Mamontova LI (1974) Meteorologicheskiy slovar (Meteorological glossary), 3rd edn. Gidrometeoizdat, Leningrad, USSR, 568p (in Russian)

Khromov SP, Petrosyants MA (2006) Meteorologiya i klimatologiya (Meteorology and climatology), 7th edn. Nauka, Moscow, 584p (in Russian)

Knippertz P, Wernli H (2010) A lagrangian climatology of tropical moisture exports to the Northern Hemispheric extratropics. J Clim 23(4):987–1003

Kondratyev KY, Timofeev YM (1970) Termicheslye zondirovaniye atmosfery so stutnikov (Thermal sounding of the atmosphere from satellites). Gidrometeoizdat, Leningrad, 410p (in Russian)

Kossin JP, Emanuel KA, Vecchi GA (2014) The poleward migration of the location of tropical cyclone maximum intensity. Nature 509:349–352

Kramer HI (2002) Observation of the Earth and its environment. Surv Mission Sens 1509p

Kutuza BG (1968) Eksperimentalnyye issledovaniya oslableniya i radioizlucheniya dozhdya v SVCh diapazone (Experimental studies of attenuation and radio emission of rain in the microwave range). Trudy GGO 222:111–120 (in Russian)

Kutuza BG, Zagorin GK, Hornbostel A, Schroth A (1998) Physical modeling of passive polarimetric microwave observations of the atmosphere with respect to the third stokes parameter. Radio Sci 33(3):677–695

Kutuza BG, Danilychev MV, Yakovlev OI (2016) Sputnikovyy monitoring Zemli: Mikrovolnovaya radiometriya atmosfery i poverhnosti (Satellite monitoring of the Earth: microwave radiometry of the atmosphere and surface). LENAND, Moscow, 336p (in Russian)

Kutuza BG et al. (2019) Pervyy v mire eksperiment po mikrovolnovomu zondirovaniyu Zemli iz kosmosa na sputnike "Kosmos-243" (The world's first experiment on microwave Earth sensing from space on the "Kosmos-243" satellite). Sovremennye Problemy Distantsionnogo Zondirovaniya Zemli iz Kosmosa 16(6):9–30 (in Russian)

Kutuza BG, Smirnov MT (1980) Vliyaniye oblachnosti na usrednennoye radioteplovoye izlucheniye sistemy "atmosfera-poverhnost okeana" (The effect of cloudiness on the averaged radiothermal radiation of the system "atmosphere—ocean surface"). Issledovaniye Zemli iz Kosmosa 3:76–83 (in Russian)

Leung LR, Qian Y (2009) Atmospheric rivers induced heavy precipitation and flooding in the western US simulated by the WRF regional climate model. Geophys Res Lett 36(3):L03820. https://doi.org/10.1029/2008gl036445

Liu WT, Tang W (2005) Estimating moisture transport over oceans using space-based observations. J Geophys Res 110(D10):D10101. https://doi.org/10.1029/2004JD005300

Lojou J-Y, Bernard R, Eymard L (1994) A simple method for testing brightness temperatures from satellite microwave radiometers. J Atmos Ocean Technology 11(2):387–400

Matrosov SY (2013) Characteristics of landfalling atmospheric rivers inferred from satellite observations over the Eastern North Pacific ocean. Mon Weather Rev 141(11):3757–3768

Mitnik LM, Mitnik ML (2003) Retrieval of atmospheric and ocean surface parameters from ADEOS-II AMSR data: comparison of errors of global and regional algorithms. Radio Sci 38(4):8065. https://doi.org/10.1029/2002RS002659

Newell RE, Newell NE, Zhu Y, Scott C (1992) Tropospheric rivers?—a pilot study. Geophys Res Lett 19(24):2401–2404

Njoku EG (1982) Passive microwave remote sensing from space—a review. Proc IEEE 70(7):728–750

Observing Systems (2011) Observing systems capacity analysis and review tool (Electronic resource). World Meteorological Organization. URL: https://www.wmo-sat.info/oscar/

Orlansky I (1975) A rational subdivision of scales for atmospheric processes. Bull Am Meteorol Society 56(5):527–530

Palmén E, Newton CW (1969) Atmospheric circulation systems: their structural and physical interpretation. Academic Press, New York, XVIII, 606P

Pan Y, Li L, Jiang X, Li G, Zhang W, Wang X, Ingersol AP (2017) Earth's changing global atmospheric energy cycle in response to climate change. Nat Commun 8(14367). https://doi.org/10.1038/ncomms14367

Peixoto JP, Oort AH (1983) The atmospheric branch of the hydrological cycle and climate. In: Street-Perrott A, Beran M, Ratcliffe R (eds) Variations in the global water budget. Springer, New York, pp 5–65

Pereslegin SV (1975) Svyaz SVCh-rasseyaniya ot morskoy poverhnosti s prostranstvenno-vremennymi harakteristikami razvitogo volneniya (Relationship of microwave scattering from the sea surface with spatio-temporal characteristics of developed waving). Izvestiya Akademii Nauk SSSR, FAO 11(5):481–490 (in Russian)

Petty GW (1990) On the response of the special sensor microwave/imager to the marine environment—implications for atmospheric parameter retrievals. A dissertation for the degree of doctor of philosophy. University of Washington, Seattle, USA, 313p

Petty GW (1994) Physical retrievals of the over-ocean rain rate from multichannel microwave imagery. Part II: algorithm implementation. Meteorol Atmos Phys 54(1–4):101–121

Pielke RA Sr (2002) Mesoscale numerical modeling, 2nd edn. Academic Press, San Diego, CA, 676p

Pielke RA Jr, Pielke RA Sr (1997) Hurricanes: their nature and impacts on society. Wiley, Chichester, 279p

Polyakov AV, Timofeev YM, Virolainen YA (2014) Using artificial neural networks in the temperature and humidity sounding of the atmosphere. Izv Atmos Ocean Phys 50(3):330–336

Ralph FM, Dettinger MD (2011) Storms, floods, and the science of atmospheric rivers. Eos 92(32):265–272

Ralph FM, Neiman PJ, Wick GA (2004) Satellite and CALJET aircraft observations of atmospheric rivers over the eastern North Pacific Ocean during the winter of 1997/98. Mon Weather Rev 132(7):1721–1745

Ralph FM, Neiman PJ, Wick GA, Gutman SI, Dettinger MD, Cayan DR, White AB (2006) Flooding on California's Russian river: role of atmospheric rivers. Geophys Res Lett 33(13):L13801. https://doi.org/10.1029/2006GL026689

Reichler T (2009) Changes in the atmospheric circulation as indicator of climate change. In: Trevor M (ed) Climate change: observed impacts on planet Earth. Elsevier, pp 145–164

Robertson FR, Bosilovich MG, Roberts JB, Reichle RH, Adler R, Ricciardully L, Berg W, Huffman GJ (2014) Consistency of estimated global water cycle variations over the satellite era. J Clim 27(16):6135–6154

Rodgers CD (2000) Inverse methods for atmospheric sounding: theory and practice. Singapore: World Scientific, Singapore, 255p

Ruprecht E (1996) Atmospheric water vapor and cloud water: an overview. Adv Space Res 18(7):5–16

Russian Hydrometeorological (2009) Rossiyskiy gidrometeorologicheskiy entsiklopedicheskiy slovar (Russian hydrometeorological encyclopedic dictionary), vol V, Bedritskiy AI (ed) Letniy Sad, St Petersburg, Moscow, pp 1–3 (in Russian)

Sasaki Y, Asanuma I, Muneyama K, Naito G, Suzuki T (1987) The dependence of sea-surface microwave emission on wind speed, frequency, incidence angle, and polarization over the frequency range from 1 to 40 GHz. IEEE Trans Geosci Remote Sens GE-25(2):138–146

Schluessel P, Emery WJ (1990) Atmospheric water vapour over oceans from SSM/I measurements. Int J Remote Sens 11(5):753–766

Sharkov EA (2000) Global tropical cyclogenesis. Springer/PRAXIS, Berlin, Heidelberg, London, NY, 361p
Sharkov EA (2003) Passive microwave remote sensing of the Earth: physical foundations. Springer-Praxis, Chichester
Sharkov EA (2005) Atmosfernyye katastrofy: Evolutsiya nauchnyh vzglyadov i rol distantsionnogo zondirovaniya (Atmospheric disasters: The evolution of scientific beliefs and the role of remote sensing). Sovremennye Sovremennye Problemy Distantsionnogo Zondirovaniya Zemli iz Kosmosa (Curr Probl Remote Sens Earth Space) 2(1):55–62
Sharkov EA (2010) Remote sensing of tropical cyclogenesis: features and scientific advances in the present state of the art. Sovremennye Sovremennye Problemy Distantsionnogo Zondirovaniya Zemli iz Kosmosa (Curr Probl Remote Sens Earth Space). 7(1):29–48 (in Russian)
Sharkov EA (2012) Global tropical cyclogenesis, 2nd edn. Springer/Praxis, Chichester, 603p
Sharkov EA, Kuzmin AV, Vedenkin NN, Jeong S, Ermakov DM, Kvitka VE, Kozlova TO, Komarova NY, Minaev PY, Park IH, Pashinov EV, Pozanenko AS, Prasolov VO, Sadovskii IN, Sazonov DS, Sterlyadkin VV, Khapin YB, Hong G, Chernenko AM (2019) Convergence space experiment: scientific objectives, onboard equipment, and methods of solving reverse problems. Izv Atmos Ocean Phys 55(9):1437–1456. https://doi.org/10.1134/s0001433819090469
Smirnov MT (1984) Modelirovaniye mikrovolnovogo teplovogo izlucheniya dozhdya metodom Monte-Karlo (Modelling the rain microwave thermal emission by Mont-Carlo method). Izvestiya Academii Nauk SSSR FAO 20(9):820–826 (in Russian)
Sohn B-J, Smith EA (2003) Explaining sources of discrepancy in SSM/I water vapor algorithms. J Clim 16(20):3229–3255
Staelin DH, Cassel AL, Kunzi KF, Pettyjohn RL, Poon RKL, Rosenkranz PW (1975) Microwave atmospheric temperature sounding: effects of clouds on the Nimbus 5 satellite data. J Atmos Sci 32(10):1970–1976
Staelin DH, Kunzi KF, Pettyjohn RL, Poon RKL, Wilcox RW (1976) Remote sensing of atmospheric water vapor and liquid water with the Nimbus 5 microwave spectrometer. J Appl Meteorol 15(11):1204–1214
Sterlyadkin VV, Pashinov EV, Kuzmin AV, Sharkov EA (2017) Differential radiothermal methods for satellite retrieval of atmospheric humidity profile. Izv Atmos Ocean Phys 53(9):979–990
Sun N, Weng F (2008) Evaluation of special sensor microwave imager/sounder (SSMIS) environmental data records. IEEE Trans Geosci Remote Sens 46(4):1006–1016
Thunis P, Bornstein R (1996) Hierarchy of mesoscale flow assumptions and equations. J Atmos Sci 53(3):380–397
Tikhonov AN, Arsenin VY (1977) Solutions of ill-posed problems. Wiley, New York, p 258
Timofeev YN (2009) Globalnaya sistema monitoringa parametrov atmosfery i proverhnosti (Global atmospheric and surface monitoring system). St Petersburg: SPbGU, 129p. (in Russian)
Troitsky AV, Gaikovich KP, Gromov VD, Kadygrov EN, Kosov AS (1993) Thermal sounding of atmospheric boundary layer in the oxygen absorption band center at 60 GHz. IEEE Trans Geosci Remote Sens 31(1):116–120
Trokhimovski YG, Bolotnikova GA, Etkin VS, Grechko SI, Kuzmin AV (1995) The dependence of S-band sea surface brightness temperature on wind vector at normal incidence. IEEE Trans Geosci Remote Sens 33(4):1085–1088
Trokhimovski Y, Kuzmin A, Pospelov M, Irisov V, Sadovsky I (2003) Laboratory polarimetric measurements of microwave emission from capillary waves. Radio Sci 38(3)(8039):4–1–4.7. https://doi.org/10.1029/2002rs002661
Ulaby FT, Moore RK, Fung AK (1981) Microwave remote sensing active and passive. Volume I: Microwave remote sensing fundamentals and radiometry. Addison-Wesley, Boston, 456p
Veltishchev NF, Stepanenko VM (2006) Mexometeorologicheskiye protsessy: Uchebnoye posobiye (Mesometeorological processes: training manual). MGU, Moscow, 127p (in Russian)
Wentz F (1995) The intercomparison of 53 SSM/I water vapor algorithms. Remote sensing systems Tech. Rep. on WetNet Water Vapor Intercomparison Project (VIP), Remote Sensing Systems, Santa Rosa, CA, 19p

Wentz F (1997) A well-calibrated ocean algorithm for special sensor microwave/imager. J Geophys Res 102(C4):8703–8718
Westwater ER, Han Y, Irisov VG, Leuskiy V, Kadygrov EN, Viazankin AS (1999) Remote sensing of boundary layer temperature profiles by a scanning 5 mm microwave radiometer and RASS: comparison experiments. J Atmos Ocean Technol 16(7):805–818
Wick GA, Neiman PJ, Ralph FM (2013) Description and validation of an automated objective technique for identification and characterization of the integrated water vapor signature of atmospheric rivers. IEEE Trans Geosci Remote Sens 51(4):2166–2176
Wilheit TT, Chang ATC (1980) An algorithm for retrieval of ocean surface and atmospheric parameters from the observations of the scanning multichannel microwave radiometer (SMMR). Radio Sci 15(3):525–544
Wu ST, Fung AK (1972) A noncoherent model for microwave emissions and backscattering from the sea surface. J Geophys Res 77(30):5917–5929
Zabolotskikh EV, Mitnik LM, Chapron B (2013) New approach for severe marine weather study using satellite passive microwave sensing. Geophys Res Lett 40(13):3347–3350
Zhu Y, Newell RE (1994) Atmospheric rivers and bombs. Geophys Res Lett 21(18):1999–2002
Zhu Y, Newell RE (1998) A proposed algorithm for moisture fluxes from atmospheric rivers. Mon Weather Rev 126(3):725–735

Chapter 3
Fundamentals of Satellite Radiothermovision

This chapter contains a description of the satellite radiothermovision approach. The combination of its constituent algorithms creates a computational scheme that is closed relative to the input information of satellite radiometry data, which provides:

(1) restoration of scalar fields of geophysical parameters of the ocean-atmosphere system of global coverage on a regular coordinate grid with a step of the order of 0.125–0.25° without data gaps and with frequent sampling (up to about 1 h) in time;
(2) restoration of the dynamics of these fields in the form of synchronous vector fields of advection (velocities of horizontal displacements);
(3) calculation of the integral characteristics of the dynamics and energy balance of the studied objects in terms of the total power of the latent heat fluxes through a given boundary or inside/outside a closed contour.

First, the physical and mathematical foundations of the proposed approach are discussed. The problem is formulated in general terms as the inverse problem of "analysis of the optical flow" and "motion estimation and compensation". In this part, the book follows the work (Ermakov 2018). The following describes the synthesis of a basic calculation scheme, partially composed of original algorithms, and partially adapting approaches known in the disciplines of technical vision and video processing, taking into account the peculiarities of the input information. This part mainly includes the results obtained in (Ermakov et al. 2011, 2019a) (the latter being the translation of original paper dated back to 2013). The next section describes the technique and the results of the analysis of the accuracy of the synthesized calculation scheme. In the most detailed form, this part of the study is presented in (Ermakov et al. 2015a), some particular aspects are mentioned in a number of other publications (Ermakov 2018; Ermakov et al. 2015b). The high accuracy of the interpolation provides the basis for advancing the basic calculation scheme in order to maximize the effective use of all available satellite radiometric information. The following

D. M. Ermakov, *Satellite Radiothermovision of Atmospheric Processes*, Springer Praxis Books,
https://doi.org/10.1007/978-3-030-57085-9_3

section is devoted to the development and implementation of an iterative multisensory approach, which relies mainly on the results of (Ermakov et al. 2016). Finally, the concluding section discusses the advantages and new possibilities of quantitative description of atmospheric processes provided by the satellite radiothermovision approach. Particular attention is paid to the calculation of energy characteristics that describe the dynamics of atmospheric processes in terms of advective fluxes of latent heat. In this part, the presentation briefly combines the results of the works (Ermakov et al. 2017, 2019b).

3.1 Physical Foundations

The tasks of reconstructing and analyzing the dynamics of the geophysical fields of the atmosphere, including the lower troposphere, were studied in the framework of the microwave radiometric research method from the very early stages of the development of satellite radiometric monitoring of the Earth (Basharinov et al. 1974). The range of issues discussed here and the analyzed dynamic characteristics of the medium is extremely wide, both in scales and in the types of physical processes studied, for example, (Gaikovich and Troitsky 1998; Gaikovich 2003; Kadygrov et al. 2003; Grankov and Milshin 2016). Further, attention is focused only on the following main aspects: (1) the possibility of calculating the short-term evolution of geophysical fields, in particular, calculating their state in a short time interval between successive observations; (2) the ability to calculate the characteristics of atmospheric movements, first of all, the advection speeds of air masses; (3) as a consequence of the first two possibilities—the ability to calculate the integral characteristics of horizontal mass and energy transfer in the atmosphere.

As noted above, early attempts to analyze the global geophysical fields of the atmosphere and their dynamics in the general context of meteorological problems were already made on the basis of the data of the first "Kosmos-243" Earth radiometric observation experiment (Obukhov et al. 1971; Basharinov et al. 1974; Gurvich and Kutuza 2010; Gorbunov and Kutuza 2018; Kutuza et al. 2019). Approaches to the statistical analysis of a set of remote data with spatiotemporal connectivity were proposed, for example (Savorskiy 1992), but the relatively small amount of information and low spatial detail were significant limitations.

One of the important components of the dynamics of the ocean-atmosphere system, of course, is the surface wind field. At the same time, the surface wind is a factor significantly affecting the thermal radiation of the ocean-atmosphere system registered from space due to changes in the state of the ocean surface. The restoration of global fields of surface wind has always been considered as one of the main tasks of satellite radiothermal remote sensing of the Earth. The possibilities of using these fields in combination with the fields of total precipitable water for analyzing the atmospheric circulation of water vapor and latent heat were also investigated. In the works (Liu 1986, 1988), this approach was implemented to obtain estimates of the monthly average advective (horizontal) fluxes of latent heat above the ocean surface, "based in part on a surprisingly good statistical relationship between integrated water vapor and surface specific humidity" (Petty 1990). It is important to note that the applied approach was based solely on data from real satellite radiothermal observations.

The development of these ideas was the synthetic approach described in (Liu and Tang 2005). The authors applied the technology of artificial neural networks to restore effective (height-integrated) values of the velocity and direction of advection of water vapor from scatterometric measurements of the speed of the surface wind and a number of additional parameters (time, geographical coordinates of the place, etc.). The obtained estimates were processed together with the total precipitable water fields reconstructed from satellite radiothermal monitoring data. As a result, despite the well-known problematic aspects, it was possible to obtain realistic estimates of the average seasonal meridional and zonal fluxes of latent heat in the latitudinal zone from 40° S up to 40° N in the four-year observation interval (1999–2003) (ibid.). A brief overview of current synthetic-type approaches is provided in (Robertson et al. 2014).

It should be noted that in all the cited works, estimates of dynamic characteristics (advection rates, heat fluxes) averaged over significant (monthly, seasonal) time intervals are considered. The detailed reproduction of short-term dynamics is hindered by the use of statistical (correlation) relationships between the sought-for characteristics and directly recovered from remote data ones. As follows from the results of (Wimmers and Velden 2011), to adequately describe the evolution of the field of total precipitable water at half-daily intervals with a resolution of the order of 10–20 km, it is necessary to know not only the surface wind field, but also the wind field at several additional horizons in the lower troposphere. However, it is impossible to extract this information from remote data based on "point" approaches that interpret measurements in each spatial viewing element (pixel) independently. This is the main feature of the problem in comparison, for example, with the task of estimating vertical heat fluxes in the ocean-atmosphere system. A fundamental solution is to use the so-called optical flow analysis algorithms (Horn and Schunck 1981; Anandan 1989; Barron et al. 1994).

As applied to the problem of analyzing atmospheric dynamics, these algorithms were used for the automated processing of geostationary satellite data. Their essence is the reconstruction of motions (air flow velocities) from changes in the spatial distributions of atmospheric moisture observed during multiple surveys in different phase states, for example, (Velden et al. 1997; Nerushev and Kramchaninova 2011). An important feature of the geostationary satellites is the ability of simultaneous observations of a significant area of the atmosphere and the Earth's surface. Such observations can be carried out with high periodicity (up to 15 min), which guarantees a detailed fixation of the evolutionary phases of even rapidly developing atmospheric processes. Unfortunately, at a geostationary orbit altitude (of the order of 36,000 km), it is technically possible at the moment to provide an acceptable spatial resolution of measurements only in the visible and IR spectral ranges. This, in turn, imposes significant restrictions on the monitoring of tropospheric processes. As a rule, using this approach, it is possible to reconstruct at each point of observation the velocity vectors of air flows at no more than two atmospheric levels simultaneously. Reliable estimates generally relate to levels above 700 hPa (or about 3000 m above the surface), while the warmest and most saturated layers of the lower troposphere lie below this level. The set of reconstructed values of the wind speed within the observation region is a mosaic, fragments of which can relate to significantly different heights, and the reliability of binding the obtained data by height is also not always satisfactory (Velden et al. 1997). This forces one to refuse to use part of the data in further analysis (for example, in assimilation within the framework of numerical models of atmospheric circulation).

Thus, this approach is not directly applicable to the analysis of the dynamic processes of the lower troposphere. However, after significant refinement and adaptation of the algorithms, the basic idea of restoring atmospheric dynamics based on the analysis of the optical flux turns out to be applicable to the data of radiothermal satellite monitoring from solar-synchronous orbits. The general concept of the satellite radiothermovision approach of atmospheric processes is as follows.

The inverse task of microwave radiometric sounding of the Earth is understood as obtaining geophysical information about the state of the studied objects based on the results of remote measurement of the characteristics of their microwave radiation (Kutuza et al. 2016; Armand and Polyakov 2005).

In the general case, this problem corresponds to the system of equations

$$T_{bi} = f_i(x_1, \ldots, x_n) + e_i, \quad i = 1, \ldots, m, \tag{3.1}$$

where T_{bi} are the measured radiation characteristics, x_1, ..., x_n are the desired geophysical parameters, e_i are the errors associated with both the measurement error and the inaccuracies of the model description f_i. Measurements are taken simultaneously in m channels.

The direct solution of system (3.1) by means of analytic inversion is not always possible and often inefficient due to the specific instability of the type of problems under consideration. In the most general form, obtaining estimates of the desired parameters can be considered as the problem of minimizing some functional

$$F(T_{b1}, \ldots, T_{bm}, \tilde{x}_1, \ldots, \tilde{x}_n) \to \min. \tag{3.2}$$

The meaning of the solution of the problem in statement (3.2) is to select such estimates for which the radiophysical model reproduces optimally, for example, according to the criterion of minimum residuals, the measured radiation characteristics or their derivatives (optical thickness of the atmosphere, etc.) (Akvilonova and Kutuza 1978; Wentz 1997).

Typically, measurements are treated as independent, point, and instant. Then the system of Eq. (3.1) can be specified in this way:

$$T_{bi}(t, \vec{r}) = f_i(x_1(t, \vec{r}), \ldots, x_n(t, \vec{r})) + e_i(t, \vec{r}), \tag{3.3}$$

where t sets the time instant of measurements, $\vec{r}$ is the point (small area) of space within the resolution spot.

However, when studying dynamic atmospheric processes, the values of geophysical parameters at time intervals and (or) in areas of space that are only partially covered by measurements, i.e. a set of values $x_j\left(t + \delta t, \vec{r} + \delta\overrightarrow{r_j}\right)$ for small deviations δt and $\delta\overrightarrow{r_j}$ from t and $\overrightarrow{r_j}$, correspondingly, are also of interest.

In this case, to use the fundamental system (3.1), it is necessary to additionally describe the relationship between the values of $x_j\left(t + \delta t, \vec{r} + \delta\overrightarrow{r_j}\right)$ and $x_j(t, \vec{r})$. By defining vectors $\overrightarrow{v_j}$ through a formal relation $\delta\overrightarrow{r_j} = \overrightarrow{v_j}\delta t$, such a relationship can be described using the simple kinematic model:

$$x_j(t, \vec{r}) = x_j\left(t + \delta t, \vec{r} + \overrightarrow{v_j}\delta t\right) + \varepsilon_j(t, \vec{r}, \delta t), \tag{3.4}$$

in which $\overrightarrow{v_j}$ can be interpreted as the velocity of the element of the field of the geophysical parameter in the linear approximation, and ε_j as a correction that includes the linear approximation error and the contribution of processes not taken into account by the kinematic model (for example, phase transitions of atmospheric moisture).

Accordingly, the parameter estimation problem (3.2) will take the form

$$\begin{gathered} F\left(T_{b1}(t, \vec{r}), \ldots, T_{bm}(t, \vec{r}), x_1\left(t + \delta t, \vec{r} + \overrightarrow{v_1}\delta t\right), \ldots, x_n \right. \\ \left. \left(t + \delta t, \vec{r} + \overrightarrow{v_n}\delta t\right), \overrightarrow{v_1}, \ldots, \overrightarrow{v_n}\right) \rightarrow \min. \end{gathered} \tag{3.5}$$

The statement of the problem in the form of (3.5) opens up the possibility for a dynamic description of the processes observed by the microwave radiometric method, but it is more complicated than in the classical interpretation of (3.2) and contains uncertainties associated with the kinematic model itself. Within the framework of satellite radiothermovision, a kind of "laminar hypothesis" was accepted that, at the considered spatiotemporal scales, the velocity fields change little in space from point to point and in the time interval between consecutive satellite measurements.

3.2 Mathematical Apparatus

3.2.1 *Optical Flow Analysis Problem*

In the most general formulation, the methods discussed below are aimed at extracting information about motion from a sequence of observations of a certain region of space. For simplicity, let the observational data be presented in the form of a *two-dimensional* dynamic field of a scalar quantity $I(x, y, t)$, the values of which are known at all cells (x, y) of a certain regular grid at time instants t repeating with a period Δt. Notice that generalization to a multidimensional case $I(\vec{r}, t)$ is quite straightforward. However, for historical reasons many of the approaches described below were specifically formulated in two-dimensional space. This appears to be satisfactory in particular when considering dynamical fields of the geophysical parameters integrated by the height of the atmosphere.

Without loss of generality, $I(x, y, t)$ can be interpreted as a dynamic image—a set of brightness values of points (x, y) at time instants t. The process that causes a change in the brightness of a certain point in the image, depending on its scale and size of the resolution spot, can either be localized at this point itself or cover a certain neighborhood of this point or the entire image area. For example, the total precipitable atmospheric water at a given location can vary both due to evaporation and condensation of water, and due to the movement (advection) of air masses.

A group of methods aimed at restoring the motion characteristics in the image plane by analyzing brightness changes of its points ("optical flow analysis") was developed as part of the tasks of technical/computer vision and video processing. The term "optical flow" emphasizes the fact that the displacements observed in the image plane have complicated and, generally speaking, ambiguous association with real three-dimensional motion. Moreover, even in the considered "flat" case, the problem of determining the displacements of image elements that lead to changes observed in the image is mathematically incorrect. For example, the image of a homogeneous circular object is invariant with respect to the angular velocity of its rotation. The choice of solution is based on a number of assumptions regarding the spatiotemporal properties of the field.

3.2.1.1 Gradient Method

As a starting point, the hypothesis is usually accepted (Horn and Schunck 1981) that the observed changes of $I(x, y, t)$ are completely determined by the displacement of its elements in the image plane:

$$I(x, y, t + dt) = I(x - u \cdot dt, y - v \cdot dt, t + dt). \tag{3.6}$$

Here dt is a small time increment; $u(x, y, t)$, $v(x, y, t)$ are the components of the sought vector velocity field of the optical flow. Under the additional assumption that the field $I(x, y, t)$ is characterized by sufficient smoothness in the time interval dt, expansion of (3.6) into a Taylor series up to terms of the second order of smallness allows relating the velocity components with $I(x, y, t)$ partial derivatives in explicit form:

$$\frac{\partial I}{\partial t} + \frac{\partial I}{\partial x} \cdot u + \frac{\partial I}{\partial y} \cdot v = 0. \tag{3.7}$$

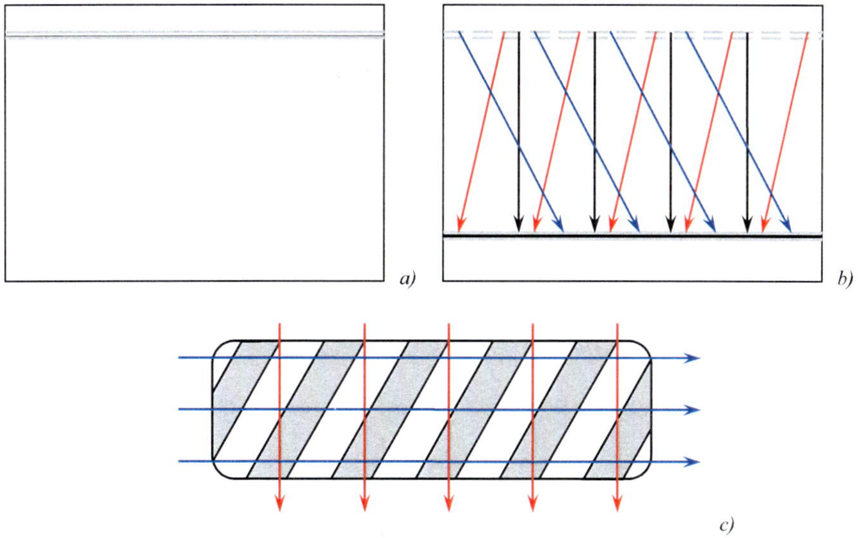

Fig. 3.1 Dimension of objects and their real and perceived movements: **a** the position of the quasi-one-dimensional object (contrast boundary) at time t; **b** the position of the border at time $t + dt$, colored arrows show some possible options for translational movement; **c** rotation of a three-dimensional object around a horizontal axis, red arrows—projections of real movement, blue arrows—direction of perceived movement

Despite the fact that the "optical flow equation" (3.7) was obtained on the basis of the strict requirement of "brightness conservation" (3.6), it was noted that it satisfactorily works in many practical applications where this requirement looks unrealistic (Nagel 1987; Barron et al. 1994; Fleet and Weiss 2005).

However, the fundamental problem is the insufficiency of one Eq. (3.7) to determine two independent components of the optical flow. It is easy to see that (3.7) determines only the projection of the flow velocity onto the direction of the $I(x, y, t)$ gradient. The displacements along the isolines of brightness are indistinguishable in the $I(x, y, t)$ field. An example of observing a fragment of a contrasting linear object moving uniformly against a uniform background is illustrated in Fig. 3.1a, b.

It is clear that between the moments of time of observation, the fragment moved down. But there is no way to determine whether it has shifted right or left during this time, and how far. Therefore, in the general case, additional assumptions are required regarding $u(x, y, t)$, $v(x, y, t)$ components and higher-order partial derivatives of $I(x, y, t)$, and a solution can be sought by expanding (3.6) into quadratic terms.

3.2.1.2 Solution Regularization

A number of approaches developed in practice lead to the emergence of an overdetermined system of equations that allows obtaining an approximate solution for the components of the optical flow velocities using $I(x, y, t)$ partial derivatives of the first and second orders. Despite some differences in the initial assumptions, at the level of final solutions, these approaches can be generalized in the model of "solid-state motion". Let the $I(x, y, t)$ gradient in the vicinity of some point substantially change direction (for example, $I(x, y, t)$ reaches a local extremum or forms a corner-shaped structure). Then both orthogonal components of motion can be restored in this neighborhood, provided that at its different points the velocity vectors are consistent with each other (in the first approximation, they are the same). This idea is formalized in (Nagel 1987) using the requirement

$$\iint_{\Omega} D(x, y, u_0, v_0, t)dxdy \rightarrow \min, \tag{3.8}$$

where Ω is a small image area in which the components of the optical flow u_0 and v_0 can be considered constant, and

$$D(x, y, u_0, v_0, t) = [I(x, y, t + dt) - I(x - u_0 \cdot dt, y - v_0 \cdot dt, t)]^2. \tag{3.9}$$

When expanding (3.9) to second-order terms, condition (3.8) gives a nonlinear system of equations that can be solved with respect to u_0 and v_0 by the method of successive approximations if in the analyzed region $I(x, y, t)$ does not degenerate into a quasi-one-dimensional structure with a constant gradient direction. Mathematically, this corresponds to the requirement that the determinant of some matrix (which differs insignificantly in different approaches) from the first and second order partial derivatives of $I(x, y, t)$ be nonzero.

Conditions (3.8), (3.9) is a kind of weakened analogue of requirement (3.6) for some image area. The resulting solution $u(x, y, t)$, $v(x, y, t)$ applies to the entire area Ω as a whole, therefore, to refine the spatial information about the movement, it is necessary to consider as small areas as possible. In this case, however, the probability increases that the $I(x, y, t)$ field will be degenerate in the Ω region in the sense indicated above. This circumstance is called the "aperture problem" in (Fleet and Weiss 2005). It should be noted that a simplified understanding of the "aperture problem" as the impossibility of reconstructing two components of the optical flow from one Eq. (3.7) is also common, which seems terminologically incorrect.

3.2.1.3 Block Methods

A group of "block" methods is aimed at finding a compromise between the detail and accuracy of the solution (Little et al. 1988; Little and Verri 1989; Anandan 1989; Grzywacz and Yuille 1990; Nam et al. 1995; Grishin et al. 2008; Liu and Ribeiro 2011), offering various approaches to optimizing the sizes and shapes of the considered areas ("blocks"), their relative position (butt, with partial overlap, with the same or varying distance between the centers of the blocks, etc.), blocks identification criteria. The most flexible are pyramidal approaches (Anandan 1989; Nam et al. 1995; Grishin et al. 2008), using a hierarchical system of blocks in the calculations. The first approximation to the solution is built on the largest spatial scale, which ensures its stability and accuracy on average due to the large size of the blocks. At each next step, using smaller blocks, the solution obtained at the previous iteration is detailed. The algorithm can be considered as an option for the implementation of the methodological approach, known under the general name "scaling" and successfully applied in a variety of tasks, including analysis of atmospheric dynamics, abnormal weather phenomena, etc. (Gledzer and Golitsyn 2010; Ruzmaikin 2014).

Requirement (3.8) can be interpreted as a criterion for the "similarity" of a block occupying a region Ω at a time moment $t + dt$ to a block shifted relative to the Ω at previous time moment t by the optical flow vector. As a metric, in addition to the sum of the squares of the differences SSD, (3.9), in practice, the sum of the absolute differences SAD is often used:

$$D(x, y, u_0, v_0, t) = |I(x, y, t + dt) - I(x - u_0 \cdot dt, y - v_0 \cdot dt, t)|. \quad (3.10)$$

The use of metrics (3.9) or (3.10) is especially attractive when there are stringent requirements on the computational efficiency of the optical flow analysis algorithm, which may be caused by the need for online processing of incoming data in real time or processing of extra-large data arrays. It is also possible to introduce the first derivatives of $I(x, y, t)$ into the metric function (Liu and Ribeiro 2011). This allows focusing the search for solutions on areas of sharp contrasts, which often correspond to the boundaries or structural features of the observed objects. The application of correlation and a number of other criteria is close in meaning (Emery et al. 1986; Liu and Ribeiro 2011), not arising directly from the above consideration, but following the general idea of finding matching blocks.

3.2.1.4 Global Methods

Further generalization of the form of metric functions allows getting rid of the requirement that the velocity of the optical flow be constant over small areas. In this case, we are no longer talking about the block method, but about the variational approach that regularizes the solution by introducing additional requirements that apply simultaneously to the entire image area. In this manner, the approach (Horn and Schunck 1981) sets a condition that weakens (3.7) while simultaneously requiring the global smoothness of the optical flow:

$$\iint \left[\left(\frac{\partial I}{\partial x} \cdot u + \frac{\partial I}{\partial y} \cdot v + \frac{\partial I}{\partial t} \right)^2 + \alpha^2 \left(\left(\frac{\partial u}{\partial x} \right)^2 + \left(\frac{\partial u}{\partial y} \right)^2 + \left(\frac{\partial v}{\partial x} \right)^2 + \left(\frac{\partial v}{\partial y} \right)^2 \right) \right] dxdy \to \min. \quad (3.11)$$

Here I, u, and v are functions of coordinates and time, α is a regularizing factor. Criterion (3.11) is aimed at searching among all (approximate) solutions of the optical flow equation of the one that has the least spatial variability. Numerous alternative optimization criteria have been proposed, for example, (Nagel and Enkelmann 1986; Weickert and Schnörr 2001; Garbe and Ommer 2013). The problem of such approaches is the known arbitrariness in choosing the smoothness criterion, which can lead to the appearance of significant artifacts (Barron et al. 1994).

The use of the above pyramidal block algorithms allows approaching the flexibility of methods of the type (Horn and Schunck 1981) while maintaining the relatively high computational efficiency characteristic of block methods and robustness to noise. It is also worth emphasizing that in all the described approaches, there is actually a rejection of the search for the exact solution (3.6), i.e., from the original model of the field of conservative tracers. This is an important circumstance for the problems of satellite studies of atmospheric processes, since phase transitions of atmospheric moisture at time intervals between successive satellite observations limit the applicability of assumption (3.6).

3.2.2 *Optical Flow Analysis in Earth Remote Sensing Problems*

Various implementations of the optical flow calculation algorithm are widely used in processing and analysis of satellite remote sensing data: for the study of sea and ocean currents (Vastano and Borders 1984; Svejkovsky 1988; Breaker et al. 1994; Bowen et al. 2002; Bobkov et al. 2003; Alexanin et al. 2011, 2013); movements and deformations of glaciers, shelf and continental ice, permafrost zones (Lucchitta and Ferguson 1986; Bindschadler and Scambos 1991; Scambos et al. 1992; Kääb and Vollmer 2000; Kääb 2002; Haug et al. 2010; Vogel et al. 2012), landslides (Yamaguchi et al. 2003), etc. An important area of application is the restoration of the so-called atmospheric motion vectors (Eigenwillig and Fischer 1982; Nieman et al. 1997; Velden et al. 1997; Menzel 2001; Key et al. 2003; Velden et al. 2005; Dworak and Key 2009; Nerushev and Kramchaninova 2011; Lazzara et al. 2014).

Atmospheric motion vectors (AMV) characterize the movement of atmospheric tracers: the characteristic boundaries of cloud structures and spatial fluctuations in atmospheric precipitable water content. Having determined the height of the tracers, it is possible to restore a fragment of the wind field on a certain horizon, considering the tracers to be passive, i.e. moving with air currents.

The calculation of AMV is carried out mainly on the basis of block methods for analyzing the optical flow (Velden et al. 1997). The approach (Nerushev and Kramchaninova 2011) is ideologically close. The analyzed $I(x, y, t)$ field characterizes the brightness temperature of the ascending atmospheric radiation at wavelengths of the mid-IR range of the spectrum (6.1–7.3 μ when analyzed by water vapor), measured from satellites in geostationary orbits. The horizontal spatial resolution in this case (about 10 km) is satisfactory for a number of practical applications (Eigenwillig and Fischer 1982; Velden et al. 1997). The advantage of this type of measurement is the high periodicity of the observations (up to 15 min), carried out simultaneously over a large area of the Earth (about 40% of the entire surface). One of the drawbacks of the approach is the lack of information on the polar regions, which they are trying to compensate by using additional measurements from low orbits. Features of the implementation of traditional algorithms lead to the fact that the field of AMV is loose (one vector characterizes a region with dimensions of the order of 32 × 32 or 24 × 24 measurement points) and is calculated on an irregular coordinate grid (blocks can be shifted in the image plane or eliminated at all for maximum solution reliability).

In the context of this book, the most serious drawback of the obtained fields of AMV is their mosaic structure. Each atmospheric motion vector is calculated for one atmospheric level characterizing the height of the observed tracer. Obtaining more than one or two vectors for a region of the size of one analyzed block is usually impossible. As a result, the reconstructed picture is fragments of the fields of the AMV at different horizons. Being certainly useful for assimilation in numerical models, it, by itself, is insufficient to restore a holistic picture of atmospheric circulation, especially in the lower troposphere. In addition, in order to calculate the latent heat fluxes in this case, it is necessary to know the vertical profile of the atmospheric moisture content for each of the regions corresponding to the analyzed blocks. Obtaining such information from satellite data with the necessary detail is an open and urgent problem of remote sensing of the Earth (Hartman 1999; Sharkov 2003; Weng et al. 2012; Kutuza et al. 2016; Sharkov et al. 2019).

3.2.3 *The Bases of the Satellite Radiothermovision Approach*

As indicated in Chap. 2, satellite radiothermal monitoring is the most reliable and informative tool for global regular monitoring of the state of the lower troposphere. The necessary horizontal spatial resolution is achieved in it due to observations from low, usually solar-synchronous orbits. This type of measurement creates additional difficulties for the application of optical flow analysis algorithms, the overcoming of which within the framework of the developed approach of satellite radiothermovision is briefly described below.

Despite the fact that in the problems of remote sensing of the atmosphere, the two-dimensional approximation (thin-film model, plane-layered model) often turns out to be adequate, a number of atmospheric processes (for example, tropical cyclones) should be considered as essentially three-dimensional. In this regard, it is necessary to pay attention to the problematic issue of the physical meaning of the restored motion fields, which often arises in vision algorithms. The differences between the "real" movements in three-dimensional space and the "perceived" movements in one of its two-dimensional projections are sometimes much deeper than trivial projective distortions. This is illustrated in Fig. 3.1c by an example of a known optical illusion.

The rotation of the Archimedes screw around its axis generates "real" movements oriented perpendicular to the axis. However, many vision algorithms reproduce the illusion of "perceived" movements along the axis of rotation, which implies that in the general case, the ratio between the direction and the magnitude of the speeds of "real" and "observed" movements can be absolutely arbitrary.

In the framework of satellite radiothermovision, there are two principal ways to solve this problem. The first, direct and most attractive, consists in the technical implementation of three-dimensional measurements, i.e., sensing the vertical structure of the atmosphere with an acceptable resolution in height. This is an extremely relevant and rapidly developing area of remote sensing. However, the results achieved here are still clearly insufficient for a practical transition to the analysis of atmospheric dynamics in a three-dimensional setting and can be used only for preliminary theoretical studies.

The second way, which has been successfully implemented in practice, consists in reducing the dimension of the problem, i.e., moving to a two-dimensional formulation: analysis of the dynamics of two-dimensional, height-integrated, geophysical fields in terms of a two-dimensional advection field. The result of the work (Wimmers and Velden 2011) can be considered as evidence of the adequacy (on the considered spatiotemporal scales) of such a two-dimensional formulation for the total precipitable water (TPW) field. The authors constructed an advection field in the form of a weighted sum of wind fields (obtained by numerical modeling) at several atmospheric horizons and showed, by comparing independent measurements from several satellites, that it describes the evolution of the TPW field in time with acceptable accuracy. The disadvantages of this approach include the fact that empirically selected weight coefficients are constant in large latitudinal zones, as if the vertical distribution of water vapor in them was everywhere standard. Despite this, the achieved accuracy of the interpolation of the TPW fields (estimated as the average discrepancy of the interpolated data with the results of independent measurements) was no worse than 2 kg/m^2 (or 2 mm in terms of the height of the column of the precipitated water), which is acceptable for practical applications (see Table 2.1 in Chap. 2).

This result can be considered a kind of "postulate of existence" of the solution. A specific way to obtain it, along with an additional improvement in accuracy and smoothness, on the basis of a calculation scheme closed with respect to satellite radiometry data, is offered by the approach of satellite radiothermovision.

3.3 Algorithms and Software Realization

When observing the Earth from polar-orbiting satellites, meridional coverage is due to the rotation of the satellite in orbit, and zonal—due to the rotation of the Earth relative to the plane of the orbit. As a result, the natural frequency of observations is ensured—12 h, if we consider independently measurements on the ascending and descending parts of the orbit—at which the coverage of the observed region is formed both on a global and regional scale. This coverage, in the general case, is incomplete:

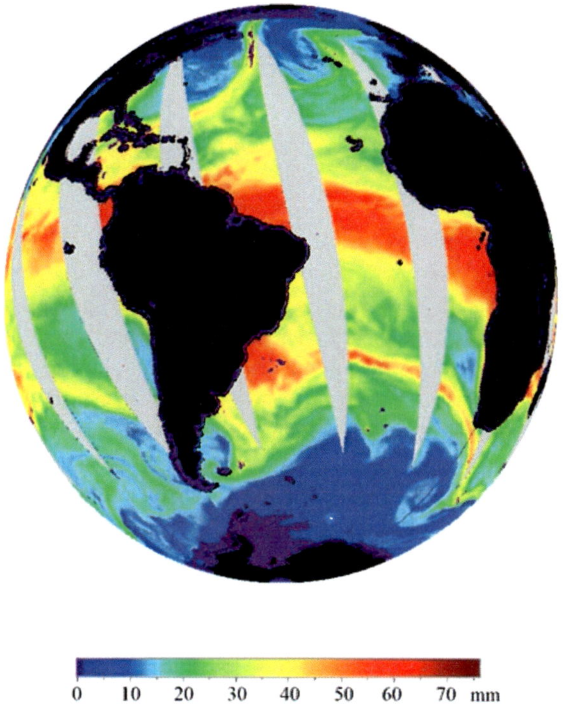

Fig. 3.2 Typical daily coverage of the Earth by satellite measurements from a sun-synchronous orbit: areas of lack of data due to the discrepancy between successive scan swaths at the equator are shown in gray, land in black. The example is built on the basis of data on the total precipitable water according to SSMIS measurements (DMSP *F*17) for 12/31/2015 (the color scale of TPW values in mm is shown below)

the divergence of the swaths at the equator leads to the appearance of spatial gaps, "lacunae"—extended sections not covered by the measurements, Fig. 3.2.

The set of measurements obtained over a twelve-hour interval, or the geophysical parameters reconstructed from them, will be called "initial fields" W_i^s. Here, the symbol W denotes a certain geophysical parameter (or measurement in the instrument channel), the upper index allows distinguishing data sources (different satellites), and the lower one allows organizing data in a chronological sequence of twelve-hour intervals. Fields W_i^s characterize the state of the observed regions at the time of measurement (we assume that all data are obtained from solar-synchronous orbits, i.e., at the fixed local time of equator crossing) and, taken separately, do not contain direct information about the dynamics of the observed processes. Such information can be obtained informally through indirect estimates, using third-party data and the experience of experts. The satellite radiothermovision approach allows formalizing and unifying the extraction of this information on the basis of a calculation scheme that is closed relative to the W_i^s data.

The basic calculation scheme consists of three main steps, the contents of which are presented below in an operator form and then described in more detail. The first step of the scheme is the formation of "reference fields" W_i, one for each twelve-hour interval for the entire set of processed data (for example, weekly, monthly, annual measurements). The absence of a superscript distinguishes the designation of the reference field W_i from the original fields W_i^s. The calculation can be represented by the action of a special operator L_R on one or more W_i^s (close in times of measurements):

$$L_R\left(W_i^{s_1}, W_i^{s_2}, \ldots\right) = W_i. \tag{3.12}$$

The essence of the operation is to combine close-in-time measurements on a single regular rectangular grid and to fill in the remaining spatial gaps with the help of smooth additional definition of W_i in the corresponding grid nodes. It was experimentally established that in the context of the tasks to be solved, measurements taken in any node within an interval of about 1 h or less can be considered sufficiently close in time to form a reference field in this node.

The reference fields W_i are ordered in chronological sequence of twelve-hour measurements in accordance with the W_i^s sequence. The need for interpolation on a regular rectangular grid and filling in the gaps is caused by the applicability conditions of the following two operators: motion estimation L_M and motion compensation L_C. The motion estimation operator L_M is applied at the second step of the calculation scheme to all pairs (W_i, W_{i+1}) of adjacent reference fields in a common chronological sequence and forms a vector displacement field $\overrightarrow{V_i}$ for each pair, which describes in a linear approximation the transformation (transition) of W_i into W_{i+1} at all nodes of the grid:

$$L_M(W_i, W_{i+1}) = \overrightarrow{V_i}. \tag{3.13}$$

The motion estimation operation is performed for each (W_i, W_{i+1}) pair iteratively on hierarchically decreasing spatial scales down to the minimum, determined by the distance between the grid nodes. As a result, macroscopic "background" motions are subsequently restored first, and then, as corrections, displacements, rotations, and deformations on ever smaller scales.

The motion compensation operator L_C is used in the third step of the calculation and forms, based on the reference field W_i or W_{i+1} or, generally speaking, their pair (W_i, W_{i+1}), and the obtained at the second step $\overrightarrow{V_i}$ field, to estimate the intermediate state $W_{i+1/2}$ of the field equidistant in time from W_i and W_{i+1}:

$$L_C\left(W_i, W_{i+1}, \overrightarrow{V_i}\right) = W_{i+1/2}. \tag{3.14}$$

The manifold of resulting intermediate fields $W_{i+1/2}$, together with the reference fields W_i, form a new chronological sequence of reference fields, to which operations (3.13) and (3.14) can be applied again. In the end, with acceptable accuracy (as shown below), the evolution can be described of the W fields with time discretization of at least 1.5 h.

Simultaneously with the spatiotemporal interpolation of the W fields, their dynamics is also estimated by constructing the displacement fields $\overrightarrow{V_i}$. After appropriate geometric calibration and normalization, these fields give the effective advection (horizontal motion) velocities observed in the W fields. This makes it possible to calculate integral physical characteristics, such as W fluxes or derived quantities, through predetermined contours with horizontal scales from 50–100 km and above.

Details of the implementation of the main steps of the basic calculation scheme are described below.

3.3.1 Constructing Reference Fields: Transformation to a Regular Grid

The motion estimation and compensation algorithm requires that the reference fields be defined on a regular rectangular grid. The grid of initial satellite measurements does not satisfy this requirement, since the scan lines represent, to a first approximation, the intersection of a sphere (surface of a geoid) with a cone (or plane) of scanning, Fig. 3.3. To represent data on regular grids of geographical coordinates, standard interpolation methods are used (Ermakov and Smirnov 2001). So, in electronic archives and databases, fields of brightness temperatures and products of their processing can be represented on regular grid, e.g., with a step of 0.5° (the so-called GRID format) or 0.25°, as, for example, in the RSS archive (Ermakov et al. 2019a).

The use of ready-to-use products brought to a regular grid has known advantages. This approach is applied in solving the problems described in Chaps. 5 and 6. However, it was found that details of 0.5° and even 0.25° are often not enough

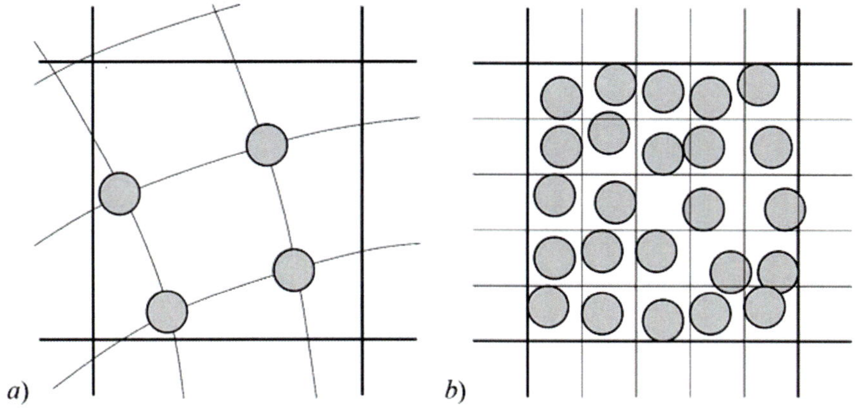

Fig. 3.3 Interpolation of SSM/I data on regular grid. Thick lines—grid cell of GRID (0.5°); circles indicate the measurements of SSM/I that fall into the GRID cell; thin lines visualize: **a** SSM/I scan lines, **b** 0.1° grid cells

to analyze the evolution of mesoscale processes. The large size of the resolution element creates two problems. The first is that field averaging within the resolution element smooths its spatial structure and makes it difficult to find and compare its characteristic features, which negatively affects the efficiency and accuracy of the optical flow analysis algorithm. The second reason is the limitation on the minimum detectable difference in the speeds of the air masses. An estimate of the sampling step in calculating the airflow velocity can be obtained by dividing the half size of the spatial resolution element by the period of its successive observations. With a grid step of 0.5° and a twelve-hour observation period, the sampling step for calculating the advection velocity at the equator will be about 2 km/h or 0.6 m/s. When using a grid of 0.125°, this estimate improves to about 0.15 m/s.

As the experience of applying the satellite radiothermovision approach to the analysis of the evolution of tropical cyclones has shown, it is advisable to carry out calculations on mesoscale atmospheric processes on a grid with a step no worse than 0.125 – 0.2°. Such detail can be provided in two ways. The first is to abandon the use of standard processing products reduced to a coarse regular grid. Instead, it is necessary to bring all the entire set of the measured radiothermal fields to a new regular grid with the required detail and then apply one of the algorithms described in Chap. 2 to calculate geophysical atmospheric fields. This method is described below. It is also possible to use the iterative scheme described in Sect. 3.5 below. Both approaches are applied in the practice of studying the evolution of tropical cyclones. The results of these studies are discussed in Chap. 4.

With a single satellite pass over the equatorial region, on average about four measurements of the SSM/I instrument fall into each cell of the "GRID" grid with a side of 0.5° (Fig. 3.3a). Accumulation in the daily measurement interval from three satellites on the ascending and descending nodes practically ensures full coverage of the nodes of the regular grid with a step of 0.1° (Fig. 3.3b). In this case, however, the asynchrony of the accumulated measurements becomes critical. As indicated above, as a result of the studies, the cell size d of 0.2° was recognized as optimal.

Given the range of acceptable latitudes (from −90° in the south to +90° in the north) and longitudes (from −180° in the west to +180° in the east), as well as a predetermined step of 0.2°, it is convenient to choose the grid width L equal 1800 cells (360°/0.2°), and height H equal 900 cells (180°/0.2°), and to denote grid cells by a pair of indices: $x = 1, \ldots, L$—the serial number of the cell in the row and $y = 1, \ldots, H$—the serial number of the row. E.g., the top row ($y = 1$) corresponds to +90° latitude (north pole), the left column ($x = 1$) corresponds to −180° longitude. The ith observation made over a point with geographic coordinates lat_i and lon_i, falls into the regular grid cell (x_i, y_i) found by the rule:

$$x_i = \left[\frac{(lon_i + 180°)}{0.2°} + 0.5\right], y_i = \left[\frac{(-lat_i + 90°)}{0.2°} + 0.5\right], \quad (3.15)$$

where the square brackets denote the operation of rounding to the nearest whole.

The value W interpolated on a regular grid will be distinguished by the presence of two lower indices, $W_{x,y}$, corresponding to the coordinates of the grid cell. In addition, we introduce an auxiliary quantity $n_{x,y}$ characterizing the filling of grid cells with interpolated values W.

Interpolation is carried out for data obtained on the daily measurement interval. Data related to measurements on different days are interpolated independently. For simplicity, consider one daily measurement interval. In fact, this means that the limits of variation of the lower index of the W_i values $i^s_{\min} \le i \le i^s_{\max}$ are chosen so that the measurement times for each of the satellites fall in one daily time interval (the upper index s emphasizes the fact that the ranges of serial numbers of measurements for different satellites can differ).

The daily data are interpolated in two stages. At the first stage, the $n_{x,y}$ values are set to zero in every cell of the grid. Then the W_i^s values are attributed to the corresponding grid cells, and the $n_{x,y}$ values for these cells set equal to 1. For this purpose, a set of data series is first selected—i.e., a sequence $\{s_1, s_2, \ldots\}$ of indices defining satellite numbers and types of orbits, see (3.12)—that will be interpolated together. Then, for each index s from this sequence, the index i runs through the entire range of its admissible values, while for each measurement W_i^s, the corresponding grid cell is calculated according to (3.10) and the values $W_{x,y}$ and $n_{x,y}$ are recalculated according to the rule:

$$W_{x,y} := \begin{cases} W_i^s, if n_{x,y} = 0, \\ W_{x,y}, if n_{x,y} = 1. \end{cases} \quad n_{x,y} := 1, i = i^s_{\min}..i^s_{\max} \tag{3.16}$$

The symbol ":=" hereinafter denotes the operation of assigning a variable a new value. As can be seen from (3.16), a change in the $W_{x,y}$ value occurs only if the corresponding grid cell has not been previously filled $\left(n_{x,y} = 0\right)$. After filling the cell, the value is set equal to 1, which eliminates the recount in (3.16).

At the end of the first stage of interpolation, a significant proportion of grid cells remains empty. This happens for several reasons. Extended areas not covered by measurements arise due to the discrepancy between the swaths near the equator or the omission of data due to temporary interruptions in instrument operation. They will be discussed in the next paragraph. In addition, there is a characteristic "moire" effect of alternating filled and unfilled grid cells, caused by a feature of the interpolation algorithm. To fill in such missing cells (taking into account the smoothness of the W field at the scale d of the grid sampling), the second interpolation step for the nearest neighboring cells is performed. The rule is adopted that interpolation in a given unfilled cell is carried out only if at least two of its neighboring cells are not empty—this cell is assigned a value equal to the arithmetic mean of the values in all filled neighboring cells. When searching for filled neighboring cells, the cyclic transition of the 180° meridian is taken into account, i.e. the leftmost and rightmost nodes of each row are also considered adjacent.

The second interpolation step is performed once, the grid cells filled at this stage are not taken into account when filling in the remaining cells.

It is important to choose a sequence of indices s used in the interpolation step according to Eq. (3.16). Attracting data from a large number of satellites or combining

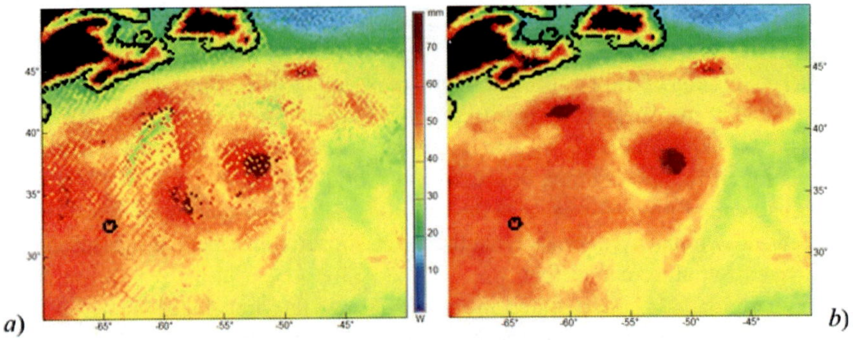

Fig. 3.4 The effect of measurement asynchrony during joint interpolation: **a** $\{s_1, s_2, s_3\} = \{F13, F14, F15\}$ DMSP satellite data; **b** $\{s_1, s_2\} = \{F14, F15\}$ DMSP satellite data. The color scale of the W values in *mm* is in the middle

data of ascending and descending orbits on the same grid leads to the appearance of characteristic artifacts related to asynchronous measurements: phantom doubling of observed objects, the appearance of sharp boundaries and shear distortions, jumps in the values of the retrieved parameters associated with the daily course of radiobrightness temperatures. Their manifestations are illustrated by an example in Fig. 3.4a, which shows an interpolated fragment of a W field obtained from measurements of satellites $F13$, $F14$ and $F15$ of the DMSP series within one day (08/12/2000). One can see the "bifurcated" body of the cyclone registered first from the $F13$ satellite, and then from the $F14$ and $F15$ satellites, as well as the characteristic "noise" due to the daily temperature variation. $F13$ data exclusion almost completely eliminates the noted artifacts (Fig. 3.4b).

3.3.2 *Constructing Reference Fields: Lacunae Stapling*

Global fields of geophysical parameters, constructed from satellites in solar-synchronous orbits, as a rule, contain "lacunae"—regions not covered by measurements due to divergence of swaths, Fig. 3.5a. Their relative area may be less than 1%, but their presence significantly complicates the application of the spatiotemporal interpolation algorithm due to edge effects at the boundaries of lacunae. Therefore, a smooth continuation of the reference fields in the region of lacunae is required, which has received the name "lacunae stapling" within the framework of satellite radiothermovision. To this end, in the vicinity of the lacunae, the directions of the smallest local W field variations (isoline directions) are calculated using a technique similar to that used in the motion estimation algorithm (Richardson 2003). Then the W field is extrapolated to the region of lacunae along the previously calculated directions (by analogy with the motion compensation algorithm). This procedure can

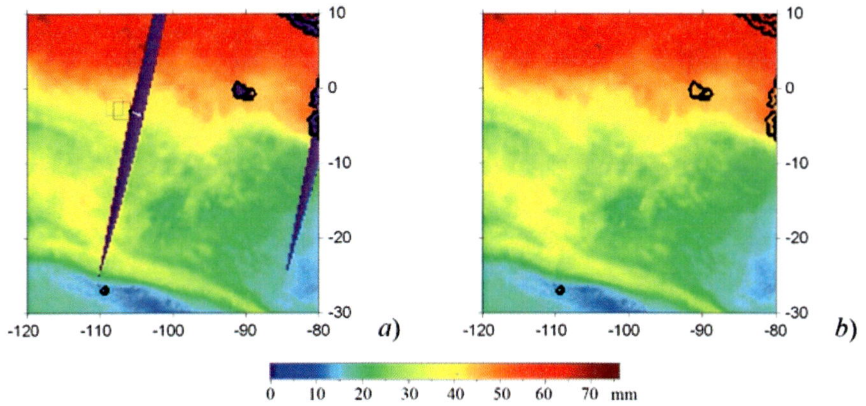

Fig. 3.5 Lacunae stapling: **a** a fragment of the TPW field, coordinates in geographical degrees; black rectangle—the analyzed neighborhood of the boundary cell of the lacuna indicated by the black circle; gray rectangle—the block closest to the analyzed area according to the SAD metric and distant at a given number of cells from the lacuna; the white arrow is the calculated continuation of the field isoline at the boundary point (extrapolation direction); **b** the result of "stapling". The color scale of the field values in *mm* is below the figure

be interpreted as a continuation of the main advective flows observed in the W field into the region of lacunae.

Such a technique does not guarantee the restoration of W values in lacunae with a certain predetermined accuracy, but is aimed at minimizing perturbations in the solution obtained by further spatiotemporal interpolation, and makes it possible to maximize the use of all available satellite data.

Next, two main steps of the lacunae stapling procedure are considered in more detail: calculation of the extrapolation direction vectors and the field extrapolation itself according to a given vector. For this purpose, an array of weighting coefficients $w_{x,y}$ defined on the entire grid is introduced. In the region of lacunae, the weighting coefficients first have a zero value and can increase by a fractional value during the stapling process. In areas filled with data after the interpolation procedure $w_{x,y} = -1$ is set.

In the following description, the rule is used to calculate the cell $\left(x^j, y^k\right)$ that is distant from the given cell (x, y) at j cells in the row ($j < 0$ corresponds to the movement to the left, to the west) and at k cells in the column ($k < 0$ corresponds to the movement up, to the north). In this case, a cyclic transition through the 180° meridian is taken into account (return to the leftmost point of the line when it reaches the right edge when moving from west to east and vice versa):

$$x^j = (x + L + j - 1)\%L + 1,\ y^k = y + k. \tag{3.17}$$

The symbol % denotes the operation of taking the remainder of integer division. It is assumed that operation (3.17) never takes the value out of the acceptable range from 1 to H, because the vicinity of the poles is not covered by SSM/I measurements and does not participate in the calculations.

The procedure for calculating the extrapolation direction vector (Fig. 3.5a) related to a given boundary cell (x, y) of the lacuna is as follows. Let the cell lie on the western border of the lacuna (gray circle in Fig. 3.5a):

$$w_{x,y} = -1,\ w_{x^{+1},y} \geq 0, \tag{3.18}$$

where the value x^{+1} is calculated according to the rule (3.17). First it is necessary to calculate the width of the lacuna at a given latitude in accordance with the following definition:

$$w_{x^l,y} = -1,\ w_{x^j,y} \geq 0,\ j = 1..l-1. \tag{3.19}$$

The width l of the lacuna determines the characteristic spatial scale d for searching the extrapolation direction vector:

$$d = \left[\frac{l}{2} - 0.5\right] \cdot 2 + 5. \tag{3.20}$$

As before, square brackets mean the operation of rounding to the integer, which is used so that the value d expressed in the number of grid cells is always odd. Thus, the minimum value taken by d is equal to $d_{\min} = 5$ grid cells, which corresponds to 1° longitude. A lower bound is necessary to ensure minimal statistics when searching for advective flows in a W field. It is noted, however, that at too large d values at the extrapolation stage, the loss of small-scale features occurs, therefore, an additional restriction is introduced from above:

$$d := \begin{cases} d, \text{ if } \ d < d_{\max} \\ d_{\max}, if \ \ d \gg d_{\max} \end{cases}. \tag{3.21}$$

The value of $d_{\max}$ is chosen equal to 19, which corresponds to 3.8° longitude. If the width of the lacuna exceeds the $d_{\max}$ value, the stapling procedure can be repeated several times, as explained below. The boundary values are established from general considerations regarding the W field structure at the observation scales and from the experience of processing a large SSM/I data array. It should be noted that the choice of fixed boundary values is not a necessary requirement of the presented methodology and can be changed when processing a specific data array.

Having determined the characteristic spatial scale d of extrapolation, the algorithm for calculating the direction of extrapolation builds to the west of the boundary cell (x, y), see (3.18), the base square window (white square in Fig. 3.5a) with side d and upper left cell $B = (x_B, y_B)$, the indices of which are determined by the rule, see (3.17):

$$x_B = x^{-d},\ y_B = y^{-\left[\frac{d}{2}-0.5\right]} \tag{3.22}$$

The algorithm for finding the direction of extrapolation uses the basic idea of the SAD criterion (Sum of Absolute Differences, the sum of the absolute differences) (Richardson 2003). The SAD criterion uses the concept of the sum of absolute differences, which for two square windows with sides d and upper left corners $B = (x_B, y_B)$ and $N = (x_N, y_N)$, accordingly, can be calculated as follows:

$$S(B, N) = \sum_{j=0}^{d-1}\sum_{i=0}^{d-1}\left|s_{i,j}\right|,$$

$$s_{i,j} = \begin{cases} W_{x_B^i, y_B^j} - W_{x_N^i, y_N^j}, if\ w_{x_B^i, y_B^j} = -1 \text{ and } w_{x_N^i, y_N^j} = -1, \\ 0, if\ w_{x_B^i, y_B^j} \geq 0 \text{ or } w_{x_N^i, y_N^j} \geq 0 \end{cases}. \quad (3.23)$$

Here, the parentheses mean the calculation of the absolute value. The value $s_{i,j}$ is introduced so that the SAD criterion takes into account only the values of W field calculated before the lacunae stapling step. In the designation of $s_{i,j}$, indications of dependence on the choice of cells B and N are omitted for compactness.

All fragments of the W field that are separated from the fragment of W in the base window are considered, see (3.22) and Fig. 3.5a, at $d/2$ to the west and distant no more than $d/2$ to the south and north (distances are measured by their upper left cells N and B, respectively). For each of the possible N sums S (3.23) are calculated. According to the original SAD criterion, the upper left cell N^* of the fragment of the W field that is least different from the W fragment in the base window (gray square in Fig. 3.5a) should be searched for by minimizing this sum.

However, the $S(B, N)$ value is affected by the number $c(B, N)$ of pairs of filled grid cells included in the calculation (3.23), which, in turn, can be variable, because depends on the shape of the lacuna in the vicinity of the base window and other factors. Accounting for this, the balanced position should be taken of all upper left cells N, with weights depending on the corresponding $S(B, N)$ values:

$$N^* = \left(x^{-\left[\frac{3d}{2}-0.5\right]}, y_S\right), p_{B,N_j} = e^{-\frac{S\left(B,N_j\right)}{c\left(B,N_j\right)d^2}},$$

$$y_S = \sum_{j=0}^{d}\left(y^{-j} \cdot p_{B,N_j}\right) / \sum_{j=0}^{d} p_{B,N_j}, N_j = \left(x^{-\left[\frac{3d}{2}-0.5\right]}, y^{-j}\right). \quad (3.24)$$

Here $S\left(B, N_j\right)$ is defined as in (3.23), $c(B, N)$ is the number of pairs of filled grid cells included in the calculation according to (3.23). The weight function p is selected experimentally in such a way as to give preference to the fragments of the W field that are least different from the fragment in the base window according to the SAD criterion, taking into account the number of cells involved in the calculation of the criterion. Now the extrapolation direction vector for a given cell (x, y) can be calculated as a vector connecting the cell N^* to the cell B, or, dividing it for further convenience by $d/2$, as follows:

$$\vec{v} = \left(v_x, v_y\right) = \left(1, \frac{2(y_B - y_S)}{d}\right). \tag{3.25}$$

The resulting vector (see Fig. 3.5a) sets the direction of extrapolation of the W field when filling the lacuna on the left (from the west). For brevity, similar equations for calculating the direction of extrapolation to the right (from the east) will not be given here. Their derivation was carried out on the basis of the same reasoning, and differences from the above expressions (3.18–3.25) are reduced to a mirror replacement of the direction of all calculations horizontally, i.e., practically, to a replacement of the sign of all variable horizontal indices and the limits of their changes by opposite ones. The second step of stapling the lacunae, i.e. the actual procedure of extrapolating the W field to the region of the lacuna in a given direction (3.25).

The extrapolation procedure consists in "propagating" the W values lying within the base window (see above) into the region of the lacuna in the direction indicated by the vector (3.25). The value W in each of the cells of the base window can contribute to several values in the nodes of the lacunae with weights that depend on the distance of "propagation". The maximum horizontal propagation distance, D, is proportional to the characteristic extrapolation scale d:

$$D = f \cdot d. \tag{3.26}$$

As a rule, the f coefficient was chosen equal to 1, however, another value less than 1, for example, 0.5 or 0.33, can be set. Such an extrapolation option may be useful for smoothing out too rugged boundaries of the lacuna and/or reducing its width until it is completely filled. In this case, the lacuna is filled after repeated stapling with the new f value.

The procedure of propagating the value $W_J = W_{x^i,y^j}$ from each filled cell $J = \left(x^i, y^j\right)$, $w_{x^i,y^j} = -1$ of the base window to the right (to the east) consists in updating the W values and weights w in the cells of the grid $\left(x^{i+m}, y^{j+n}\right)$ through which passes the vector (3.25) starting in the cell J and having the length of the horizontal component equal to (3.26). Updating W is carried out by adding a term proportional to the starting value of W_J and the "attenuation coefficient" determined by the current propagation distance.

$$W_{x^{i+m},y^{j+n}} := \begin{cases} W_{x^{i+m},y^{j+n}} + W_J \cdot e^{-\frac{2m}{d}}, \text{if } w_{x^{i+m},y^{j+n}} \geq 0, \\ W_{x^{i+m},y^{j+n}}, \text{if } w_{x^{i+m},y^{j+n}} = -1. \end{cases}$$

$$w_{x^{i+m},y^{j+n}} := \begin{cases} w_{x^{i+m},y^{j+n}} + e^{-\frac{2m}{d}}, \text{if } w_{x^{i+m},y^{j+n}} \geq 0, \\ -1, \text{if } w_{x^{i+m},y^{j+n}} = -1 \end{cases}$$

$$n = \left[m \cdot v_y\right], m = 1..D. \tag{3.27}$$

As can be seen from (3.27), those W values at the grid cells with negative weights, i.e. calculated before the current stitching procedure, for example, in the interpolation step, cannot be recalculated. The indices m and n vary in a consistent manner so that the direction of propagation of the value from the starting cell corresponds to the vector (3.25). The limit on the maximum distribution is set by the limit m value (horizontal shift) equal to D. Procedure (3.27) is performed for all admissible values of m for each of the filled cells J within the base window. Similar equations for propagating W values to the west from the eastern vicinity of the lacuna will not be given for brevity.

The algorithm searches for nodes that belong to the western and eastern boundaries of the gaps, sequentially passing through all the nodes of the grid, and for each found boundary node it calculates the extrapolation vector (3.25) and carries out the procedure of propagating the W values. Only after these procedures are performed for all boundary cells, the values $W_{x,y}$ in the lacunae are normalized taking into account the accumulated weights $w_{x,y}$:

$$W_{x,y} := \begin{cases} W_{x,y}/w_{x,y}, \text{ if } w_{x,y} > 0, \\ W_{x,y}, \text{ if } w_{x,y} \le 0. \end{cases} \tag{3.28}$$

The stapling result for a fragment of the field W shown in Fig. 3.5a is shown in Fig. 3.5b. If after a single stapling some lacunae remain not completely filled, the procedure can be repeated. To do this, it is enough to re-initialize the array of weights $w_{x,y}$, equating them to -1 at all cells where they have nonzero values. The meaning of initialization is to set all the W values calculated by this moment (interpolated and extrapolated) as the source (unchanged) for the next stapling procedure.

3.3.3 Spatiotemporal Interpolation

Further spatiotemporal interpolation of the reference fields in basic terms reproduces the classical algorithmic solutions to the problem of motion estimation and compensation. The motion estimation step (reconstruction of the motion vector field) is implemented using the pyramidal block method of optical flow analysis (Anandan 1989). Ideologically, it is very close to the algorithm for calculating the local direction of the field isoline (3.25) constructed above, with the significant difference that the search for correspondence by a block is carried out not for one reference field, but for a pair of reference fields corresponding to two consecutive observations. Similarly, the motion compensation step is close to the algorithm for extrapolating the field into the lacuna (3.26–3.28). For these reasons, a detailed description of these algorithmic steps is not required. Attention is paid only to some qualitative aspects of the implementation of the calculation scheme.

Figure 3.6 schematically illustrates the motion estimation step performed using the pyramidal block method of optical flow analysis. The algorithm uses two reference fields, $W(t)$ and $W(t + \Delta T)$ chronologically following each other (t is the conditional moment of time of observations; ΔT is the period of the observations). The $W(t)$ field is divided into N blocks $B_i^{(1)}$, $i = 1..N$ of geometrically equal regions

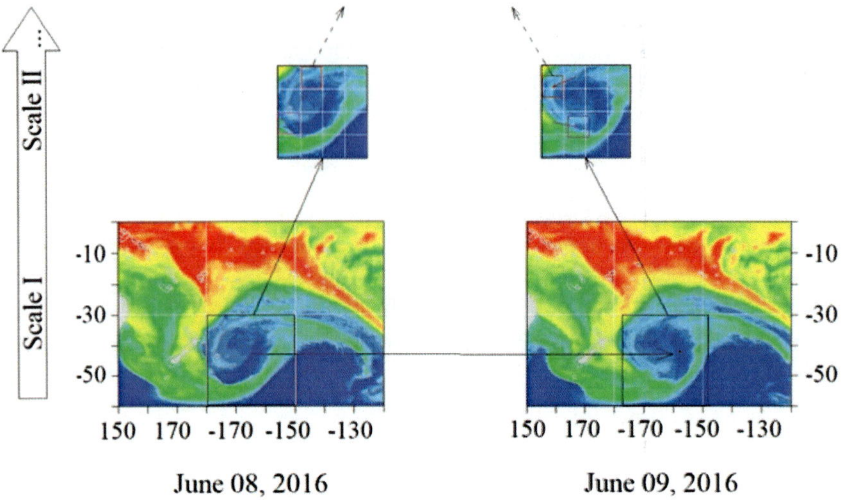

Fig. 3.6 Schematic illustration of the pyramidal method of motion estimation. The images on the left side are the total precipitable water values according to satellite radiometry data for June 8, 2016 (local time about 15:00); on the right side—for June 09, 2016 (about 03:00). The bottom row of images is analysis on the largest spatial scale. A displacement of the block bounded by a black frame by approximately 2.5° to the east was revealed. The top row of images is an analysis of the corresponding blocks on a more detailed spatial scale. The displacements of individual blocks are revealed, shown in the upper right image by arrows. Further, the analysis continues on successively decreasing spatial scales

that collectively overlap the entire computational grid (rectangles in Fig. 3.6), i.e. so that each grid cell is contained in at least one of the blocks. The position of the *i*th block is uniquely determined by a pair of numbers (x_i, y_i) that specify the coordinates of a particular (for example, angular) block cell. For each block $B_i^{(1)}$ of the reference field $W(t)$, a search for "similar", i.e. closest by a given metric block $B_i^{\prime(1)}$ in the field $W(t + \Delta T)$ is performed. The search is carried out in a limited (proportional to the size of the block) neighborhood of the cell (x_i, y_i). If the most similar block is found at position $\left(x_i', y_i'\right)$, then the *i*th block is associated with the displacement vector $\overrightarrow{m_i} = \left(x_i' - x_i, y_i' - y_i\right)$.

As a result of applying this procedure to all blocks of the reference frame, a vector field $\vec{M}$ is formed connecting pairs of $W(t)$ blocks with similar and adjacent $W(t + \Delta T)$ blocks. The displacement vectors forming the field $\vec{M}$ do not always make sense of object motions in the observation plane. The proximity of a pair of blocks in a given metric does not mean the identity of the observed objects, especially if the best match found is not "good enough". In particular, the problem of choosing a false target from equal (by the criterion of proximity) alternatives may arise. The motion estimation algorithm must satisfy the requirements of continuity of the obtained displacement field and robustness to mutual displacements (or deformations) of objects of different scales. For this purpose, the procedure is built hierarchically. First, an analysis is carried out for large blocks, giving a rough estimate of the field at the largest scales, then the sizes of blocks and search regions are successively reduced (and the centers of the search regions are shifted in accordance with the previously obtained estimates of the displacement vectors), providing an ever more detailed approximation while maintaining $\vec{M}$ smoothness.

For the same purpose, to obtain a more detailed and accurate idea of the $\vec{M}$ field, a number of other methods are used. Thus, the blocks of the reference frame are placed not "in joint" (so that each pixel of the frame belongs to exactly one block), but overlapping each other, reducing the distance between the centers of the blocks up to one pixel in each direction. The lengths of the displacement vectors are calculated with subpixel accuracy, having previously carried out the simplest (for example, bilinear) interpolation of the compared frames. If there is a priori information about the anisotropy of the observed objects or the preferred direction of their movement, the sizes of the blocks and/or the search area can vary independently in different dimensions.

Assuming that the $\vec{M}$ field adequately describes the displacements of objects (W field elements) over a known time interval ΔT, it is easy to calculate (in a linear approximation) the velocity field of the observed objects and evaluate the positions of these objects at intermediate points in time, i.e., in fact, perform spatiotemporal interpolation. This procedure is sometimes called "motion compensation" (Richardson 2003; Grishin et al. 2008). For this, it is necessary to implement the operation of the action of the specified displacement field (the $\vec{M}$ field multiplied by a coefficient k in the range from 0 to 1) on the reference field $W(t)$, the result of which is a new field defined on the same computational grid $W(t + k\Delta T)$, with correspondingly displaced elements (blocks). As indicated above, this operation is completely similar to the calculation steps described above (3.26–3.28). One of the difficulties is that the blocks of the reference field after the displacement will not necessarily collectively cover the entire computational grid. To eliminate the gaps, the displacement field should be smoothly detailed by calculating the average displacement vectors for the blocks covering the neighborhood of the data gap, and finding (or interpolating) the region of the reference field that closes this gap as a result of the calculated displacement. The most acute problem arises at the edges of the computational grid. In this case, the method of closing gaps described above may not work, and the temporal interpolation of such areas must be carried out by more primitive means, for example, by the method of local "optimal" time interpolation of the processed pair of fields. In this regard, to work with global TPW fields, the computational grid is constructed so that its left (western) and right (eastern) edges extend mainly over land (20° E).

Since during the time interval between a pair of interpolated geophysical fields the observed objects can not only move, but also change their local properties (for example, due to phase transitions of atmospheric moisture), then to improve accuracy, the motion compensation procedure should be implemented symmetrically with respect to both frames. In this case, to calculate the state of the field at a point in time $t + k \cdot \Delta T$, the displacement vector field is acted upon $W(t)$ by the vector displacement field $k \cdot \vec{M}$, and the field $(k - 1) \cdot \vec{M}$ acts on the reference field $W(t + \Delta T)$, after which the results of the two actions are combined by analogy with the above algorithm (3.26–3.28).

The field interpolated at an intermediate moment of time $W(t + k\Delta T)$ is, in the end, a weighted sum of the transformed fragments of the reference fields and, in the general case, is not identical to the translation transformation of any of them using some effective vector displacement field $\overrightarrow{M'}$. For this reason, in order to achieve a high discreteness of time interpolation (reduction of the step $k\Delta T$), the approach presented here does not perform with the same $\overrightarrow{M}$ field for different values of k. Instead, a value $k = 0.5$ is fixed, which corresponds to interpolation at the moment equidistant in time from two consecutive observations. After performing the calculations, a new sequence of reference fields is formed, where the interpolated ones in the previous step are included, after which the motion estimation and compensation steps are repeated. Each repetition leads to an effective decrease in the time sampling step by a factor of two, i.e. from 12 h to 6, 3 and 1.5 h, respectively. As a rule, a further decrease in the sampling step does not make sense, since the initial satellite data used in constructing the reference fields can be considered "synchronous" over the same observation areas with an accuracy of about 1–1.5 h. An exception may be limited data samples used for regional analysis of mesoscale atmospheric processes.

Together, the algorithms described in the previous paragraphs provide the possibility of spatiotemporal interpolation of geophysical fields, taking into account the dynamics of multiscale processes occurring in them. In addition to restoring the state of the fields to time instants between the measurements, vector displacement fields are calculated that characterize, up to geometric correction and normalization, the average velocities of advection (horizontal movements) in the observed fields. Moreover, as one can see, there are many options for implementing and tuning the interpolation scheme, and a priori estimation of its accuracy is almost impossible. The construction of a specific implementation, the most adequate to the goal of the study, requires an empirical selection of parameters. Then an a posteriori estimation of the interpolation accuracy is carried out. The results of the analysis of the accuracy of the calculation schemes used in the framework of the topics presented here are discussed in the next section. The analysis was carried out according to a technique completely similar to that used in (Wimmers and Velden 2011) for assessing the accuracy of the "seamless advective blending" method proposed there.

3.3.4 Some Brief Notes on Software Implementation

Several conceptually close software solutions were realized in several ways during the research which results are discussed in the book. The "lacunae stapling" algorithm was implemented as a console utility (Ermakov et al. 2011), integrated, among others, in the software of the geoportal of satellite radiothermovision (see Chap. 7). It is in many aspects pretty close to the motion estimation/compensation algorithm using pyramidal approach. The software implementation of the latter was, e.g., performed and demonstrated by the author and colleagues (Ermakov 2017) on the base of the FileX EZ-Vizi (the former Stream Handler) platform (FileX 2000). The reader is also recommended to study an open-source project AviSynth (AviSynth 2002) which provides a lot of powerful state-of-art tools and useful guidelines of image and video processing including different similar approaches to spatiotemporal image/video interpolation.

3.4 Accuracy Analysis

In many cases, the global diurnal fields of a certain geophysical parameter W can be obtained from at least two independent series of satellite microwave measurements, A and B. We will denote such "initial" fields W^A and W^B and assume that they are built on a regular geographic grid, as described in the previous paragraphs. The values of the fields W^A and W^B in the same grid cell i are denoted as W_i^A and W_i^B respectively.

For a given day of observation, the difference

$$\delta_i = W_i^A - W_i^B \tag{3.29}$$

will generally be nonzero for two main reasons. Firstly, because of the errors in reconstructing the W fields and individual characteristics of measurements by different instruments, and secondly, because of the natural changes that occurred in the W field in the time interval Δt between measurements by these devices. In the case of solar-synchronous orbits, the time Δt is approximately constant for all grid cells and is determined by the local time of the ascending (descending) nodes of the two satellites. Satellite radiothermovision allows for spatiotemporal W field interpolation (for example, based on the source data W^A) and, thus, for each node i select the value $W_i^A(t)$ that is maximally "synchronized" with the corresponding W_i^B value. In this case, the absolute value of the difference (3.24) should decrease due to minimization of Δt. However, additional interpolation errors arise, which can affect the increase in the difference (3.29).

As an integral criterion for the difference between the interpolated field $W_i^A(t)$ and the independently obtained W_i^B at a fixed Δt value, the average absolute value of δ_i for all filled cells of the grid can be used (and more generally for all N measurements of a long series):

$$e_\delta(\Delta t) = \frac{1}{N}\sum_{i=1}^{N}|\delta_i|. \tag{3.30}$$

The study of quantity e_δ as a function of Δt allows analyzing the "synchronizing" effect of spatiotemporal data interpolation, i.e. expected decrease of e_δ as the absolute values of Δt decreases. At the same time, the minimum achievable e_δ value is a natural measure of the accuracy of spatiotemporal interpolation. A similar approach and the same criterion for the accuracy of interpolation were adopted in (Wimmers and

Velden 2011) and used here for ease of comparison. In addition, the approximation of the obtained δ_i distributions for various Δt is presented using Gaussian function

$$g(x) = Ae^{-\frac{(x-\mu)^2}{2\sigma^2}} \tag{3.31}$$

and Cauchy-Lorenz function

$$f(x) = A\frac{\gamma}{(x-\mu)^2 + \gamma^2} \tag{3.32}$$

to assess their symmetry, the presence of systematic error and the probability of occurrence of significant interpolation errors.

As the main data series in the study of the accuracy of the interpolation scheme, standard products of Remote Sensing Systems (RSS) were used, namely daily fields of the total precipitable water, reconstructed from measurements by SSMIS instruments from $F16$ $\left(W^{F16}\right)$ and $F17$ $\left(W^{F17}\right)$ satellites of the DMSP series in November 2013. Millimeters (the effective height of a column of condensed water) are taken as a unit of measurement for the values W and, respectively, residuals (3.29) and average interpolation errors (3.30), 1 mm corresponds to a moisture content of 1 kg/m^2. Spatiotemporal interpolation was performed on products W^{F16} supplemented by insignificant amounts of other data for the implementation of the procedure for constructing reference fields (see above). In addition, the data processing products of AMSR-2 $\left(W^{A2}\right)$ and SSM/I $F15$ $\left(W^{F15}\right)$ were used in the analysis. Details of the processing and analysis procedures are described in (Ermakov et al. 2015a).

When comparing the initial W^{A2} fields (according to AMSR-2), unlike the W^{F17} fields, with the interpolated W^{F16} fields, it is necessary to take into account the presence of additional factors that increase the error (3.30) and cannot be eliminated by interpolation: differences in the characteristics of measuring instruments, orbit parameters, and calibration and restoration methods of W values. In this part of the study, we were primarily interested in: the average error (3.30) for the non-interpolated SSMIS and AMSR-2 data (since in complex studies the data of various satellite devices are often used together without preliminary interpolation); qualitative effect of interpolation, i.e. error change (3.30) as data synchronizes; as well as the magnitude of the error achieved with the best synchronization.

In all cases of comparison, we used data samples of the order of 10^7 pairs of interpolated and independently measured (reconstructed based on satellite radiometric measurements) W values on the global grid with a step of 0.25°.

Figure 3.7 shows the results of calculating the interpolation error e_δ as a function of Δt when comparing the interpolated W^{F16} fields and the non-interpolated W^{F17} fields. Negative Δt values correspond to cases of interpolation of $F16$ data at earlier times than the measurement time of $F17$ at the corresponding grid cells. The best synchronization of data series is provided with a $\Delta t = +0.5$ h value. The $\Delta t = -1$ h value, indicated by a vertical arrow in Fig. 3.7, corresponds to the time interval between the real (closest in time) measurements $F16$ and $F17$.

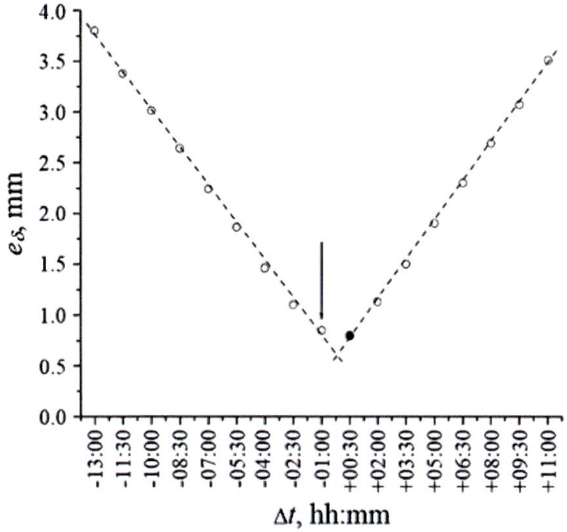

Fig. 3.7 The average residual of the interpolated TPW fields according to SSMIS $F16$ with independent measurements of SSMIS $F17$ as a function of "synchronization"

The following main features of e_δ should be noted. Firstly, e_δ as a function of Δt demonstrates a near-linear decrease simultaneously with a decrease in the absolute value of Δt. In the considered interval of ± 12 h, the values of e_δ lie in the range of 0.8–4.0 mm. Both of these features are in good agreement with the results obtained in (Wimmers and Velden 2011) and, in particular, shown in the plot of Fig. 3.8 of this work (not reproduced here).

In addition, the minimum e_δ value is achieved with the minimum absolute value of Δt, and straight lines approximating two groups of e_δ values for positive and negative Δt converge at Δt close to 0 (dashed lines in Fig. 3.7). This confirms the

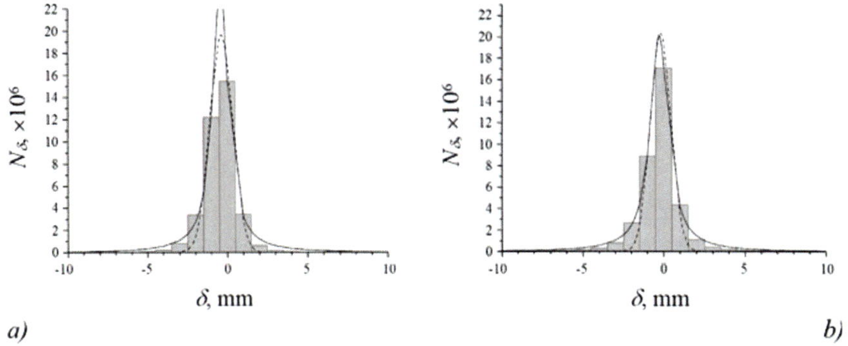

Fig. 3.8 Distribution of interpolation residuals for cases **a** $\Delta t = +0.5$ h and **b** $\Delta t = -1$ h; approximation by the functions of Gauss (dotted line) and Cauchy-Lorentz (solid line)

adequacy of the description of the W field evolution by the adopted interpolation scheme. An integral measure of the error e_δ of the adopted interpolation scheme should be considered the value calculated for the minimum absolute Δt. This value in the studied data sample was 0.8 mm (Fig. 3.7, black circle), which lies within the interval of 0.5–2.0 mm, and is very close to the maximum achievable value of 0.5 mm, determined by errors in the initial data. In particular, it is smaller than the same value calculated for the non-interpolated data $F16$ and $F17$ (0.85 mm), and also less than the errors of 1.0–1.2 mm of the scheme from (Wimmers and Velden 2011) arising from extrapolation of the source data to the minimum time interval considered forward or backward considered in the cited work (Fig. 3.8 of this work).

Figure 3.8 shows the distribution of the entire set of residuals δ_i (3.24) between the initial W^{F17} fields and the interpolated W^{F16} fields for cases a) $\Delta t = +0.5$ h (W^{F16} fields are maximally "synchronized" with W^{F17}) and b) $\Delta t = -1$ h (non-interpolated W^{F16} fields) in the form of a histogram of the number of cases N_δ in which the residual amounted to value δ mm. The distributions have a pronounced monomodal character and are almost symmetrical. They can be approximated with high accuracy by the functions of Gauss (3.31) and Cauchy-Lorentz (3.32). The corresponding approximation parameters can be considered as an estimate of the systematic (μ) and random (σ, γ) interpolation errors, bearing in mind that the initial reconstruction errors of W also contribute to them.

The proximity of the average value (shear parameter) to 0 and, in particular, the small differences of these values (about 0.2 mm) for the cases considered indicate the absence of a significant systematic interpolation error. The narrowness of the distribution (the σ value of the rms spread, scale parameter γ) once again shows the overall high accuracy of the interpolation. Based on the obtained distribution parameters (3.31), (3.32), we can estimate the probability of the occurrence of significant residuals (3.29) at individual cells of the computational grid. Table 3.1 summarizes the results of the analysis: values of the mean (shift parameter) μ, standard deviation σ (scale parameter γ), and approximation quality criteria R^2. In addition, using approximations of the distribution by functions (3.31), (3.32) on the $|\delta_i| \leq 75$ mm segment, the quantities $\delta_{0.95}$ and $\delta_{0.99}$, which do not exceed the absolute value of the residual in 95% (99%) cases (analogues of distribution quantiles) for the $\Delta t = +0.5$ and $\Delta t = -1$ h, are numerically estimated. The obtained values (see Table 3.1), when approximated by the Cauchy-Lorentz function (3.32), lie within the corresponding ranges of values obtained in (Wimmers and Velden 2011): 3–7 mm errors for 95% of

Table 3.1 Distribution characteristics of interpolation residuals

Δt, h	Approximation by the Gauss function (3.26)					Approximation by the Cauchy-Lorentz function (3.27)				
	μ, mm	σ, mm	R^2	$\delta_{0.95}$, mm	$\delta_{0.99}$, mm	μ, mm	γ, mm	R^2	$\delta_{0.95}$, mm	$\delta_{0.99}$, mm
+0.5	−0.39	0.83	0.994	1.7	2.6	−0.44	0.56	0.996	4.2	8.6
−1.0	−0.18	0.74	0.986	1.5	2.4	−0.27	0.64	0.998	4.7	9.5

cases and 5–12 mm for 99% cases, and when approximated, the Gaussian function even turns out to be much lower.

It should be noted that the Cauchy-Lorentz function (3.32) describes the δ_i distribution edges somewhat better (see Fig. 3.8 and Table 3.1). A possible explanation is that as a result of preprocessing the data in RSS, including complex calibration procedures, the values of the residuals δ_i (differences between W^{F16} and W^{F17}) cease to be normally distributed values, and instead the value depending on the ratio of W^{F16} and W^{F17} takes on the character of the normal distribution. The indicated δ_i distribution properties are preserved at Δt values different from those considered. In this case, the distribution width only increases. To illustrate this fact, Table 3.2 shows the corresponding values of the shift parameter μ (position of maximum), scale parameter γ (measure of width), and approximation quality (R^2 criterion) of the Cauchy-Lorentz distribution (3.32) for a number of Δt values. The table shows that the γ parameter ("half-width" of the distribution) monotonically decreases to 0.56 mm when Δt approaching 0.5 h value on both sides. In this case, the distribution center (μ parameter) varies in a narrow range of values from -0.44 to -0.16 mm (-0.27 mm for non-interpolated data). There is no significant contribution of interpolation errors to the systematic error of W reconstruction.

Figure 3.9 shows the results of calculating the interpolation error e_δ as a function of Δt when comparing the W^{F16} fields (interpolated) and W^{A2} fields (non-interpolated). Negative Δt values correspond to cases of interpolation of SSMIS data at earlier times than the time of measurement of AMSR-2 at the corresponding cells of the grid. The best synchronization of data series is provided with the $\Delta t = +0.5$ h value. The $\Delta t = +3.5$ h value, indicated by a vertical arrow in Fig. 3.9, corresponds to the time interval between real measurements of SSMIS $F16$ and AMSR-2.

Figure 3.9 reproduces all the main features of the graph in Fig. 3.7. The only significant difference is the increased value of the minimum of e_δ achieved with the best field synchronization. The nature of the e_δ dependence on Δt suggests that the magnitude of the error is not due to the accumulation of interpolation errors, but, first of all, the heterogeneity of the origin of the data series (different instruments, observation parameters, and processing algorithms) and the differences in the W estimates associated with it. In any case, spatiotemporal interpolation allows reducing the $e_\delta = 2.5$ mm value obtained for the initial (not interpolated) data to 2.0 mm with the best "synchronization", which falls within the range of 0.5–2.0 mm. It should be noted that in (Wimmers and Velden 2011) the indicated error range was obtained by analyzing only SSMIS $F16$ and $F17$ data. A more detailed description of the analysis is contained in (Ermakov et al. 2015a).

In conclusion of this section, we should also briefly dwell on the issue of the accuracy of reconstruction of advection vector fields calculated at the step of motion estimation, which is methodologically more complicated (Ermakov 2018). Independent experimental data in a statistically significant volume, in fact, are absent. Comparison with semi-empirical estimates based on modeling data, for example, (Wimmers and Velden 2011), raises the counter question of the accuracy of the latter. Indirect evidence of the generally satisfactory accuracy of the restoration of the advection field is the high accuracy of the restoration of the interpolated TPW

Table 3.2 Parameters for approximating the distribution of residuals by the Cauchy-Lorentz function

Δt, h	−8.5	−7.0	−5.5	−4.0	−2.5	−1.0	+0.5	+2.0	+3.5	+5.0	+6.5	+8.0
μ, mm	−0.37	−0.37	−0.39	−0.39	−0.43	−0.27	−0.44	−0.35	−0.31	−0.22	−0.22	−0.16
γ, mm	1.83	1.55	1.29	1.05	0.80	0.64	0.56	0.89	1.13	1.39	1.66	1.94
R^2	0.994	0.996	0.995	0.996	0.996	0.998	0.996	0.997	0.997	0.998	0.997	0.997

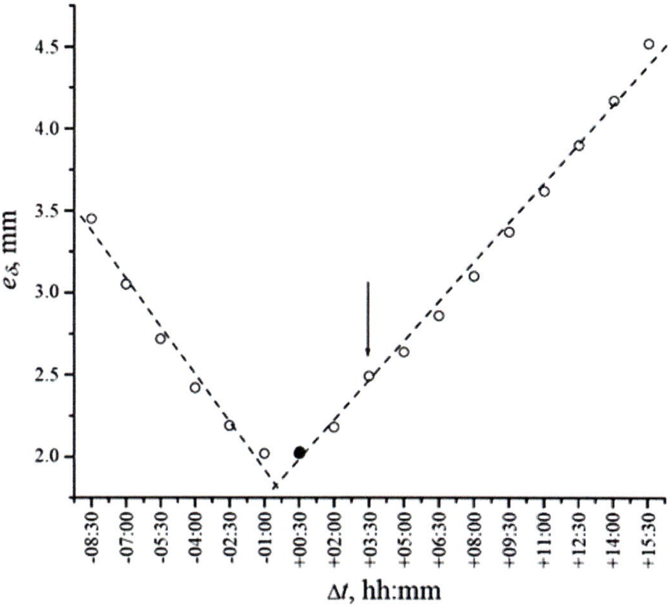

Fig. 3.9 The average residual of the interpolated TPW fields according to SSMIS *F*16 with independent measurements of AMSR-2 as a function of "synchronization"

fields (see above). Unfortunately, such a check, in principle, does not exclude the possibility of a systematic error that may occur at the step of motion estimation and be "self-consistently" compensated at the step of motion compensation. To eliminate this possibility, test processing of specially synthesized sequences of fields shifted relative to the initial sample by a given constant vector was carried out. The analysis results showed that in this simplest case, the average error in the reconstruction of the displacement vector is no worse than one grid cell for each measurement in a wide range of advection vector lengths (up to 64 cells or 1600 km with a grid sampling of 0.25°) (Ermakov 2018). The main contribution to the average error, as one would expect, is made by the areas of sharp discontinuity: the edges of the grid, the land – water transition boundaries. At low advection rates, the spatiotemporal discretization of the source data can become a significant error factor. Given that the motion estimation algorithm works with subpixel resolution (Ermakov et al. 2011), the movement becomes completely indistinguishable when the elements of the field pass in 12 h less than half the distance between adjacent grid cells, i.e. moving at a speed of less than 1.16 km/h or 0.32 m/s. Such discretization seems quite satisfactory for the tasks included in the consideration here.

An additional analysis of the accuracy of restoration of advection fields was carried out by comparing the dynamics of the positions of the radiothermal images of tropical cyclones in the interpolated fields with their positions known from independent measurements. Most clearly and fully, such a comparison is possible with visualization in the streaming video mode (Zhu and Newell 1994; Ermakov et al. 2011). Figure 3.10 shows an illustrative example, which can be considered as a sequence of frames of a short fragment of such a video.

In the given example, "instantaneous" (that is, interpolated to a fixed moment of universal time, UTC) global TPW fields are used. The methodology for constructing "instant fields" is described below in paragraph 3.5 and is more detailed in (Ermakov et al. 2016). Symbols denoting the positions of tropical cyclones in the hurricane (typhoon) stage are superimposed on these fields. Their image in the TPW field is a clearly defined spiral, which allows determining the position of their centers with an error of several cells of the computational grid, i.e. fraction of a degree. Such accuracy is quite satisfactory for solving the problems described in subsequent chapters of the book. Independent information on the trajectories and other characteristics of tropical cyclones was obtained from (Pokrovskaya and Sharkov 2006).

It should be emphasized that the positions of the centers of tropical cyclones in the TPW field were determined manually, and the analysis should be considered first of all from the point of view of adequacy of reproducing the magnitude and direction of displacements of small elements of interpolated fields, and not the problem of determining the absolute positions of the centers of tropical cyclones (effectively solved by other means of remote sensing).

The example in Fig. 3.10 illustrates the state of the atmosphere on 08/28/2005. The TPW fields are interpolated at time points (a) 12:00 UTC; (b) 15:00 UTC; (c) 18:00 UTC; (d) 21:00 UTC. For this time interval, the database (Pokrovskaya and Sharkov 2006) contains information on two tropical cyclones (TC) that have reached the stage of a hurricane (typhoon). These are the TC Katrina in the North Atlantic and the TC Talim in the Northwest Pacific. Data on the position of the center of the TC Katrina is available at 12:00, 15:00 and 21:00 UTC, for the TC Talim—at 12:00 and 18:00 UTC (i.e. with a time step of no worse than 6 h). In the corresponding images they are shown by white circles. For clarity, coordinates of the centers of the TC Katrina at 18:00 UTC and the TC Talim at 15:00 UTC and 21:00 UTC are interpolated from neighboring values and are also shown in white diamonds on the corresponding images.

It can be seen that the interpolated TPW fields adequately reflect the dynamics of movement of both tropical cyclones. It should be noted that for the considered time interval, the TC Katrina moved 1.2° in latitude and 1.3° in longitude (about 5 cells of the computational grid in each direction), moving at an average speed of about 5 m/s. TC Talim moved approximately 0.9° in latitude (about 3.5 cells of the computational grid) and 1.9° in longitude (about 8 cells of the computational grid), respectively, moving at an average speed of about 6 m/s.

In general, the average discrepancy between the positions of the centers determined in the TPW fields and known from independent sources in the analyzed examples does not exceed 0.5° (the angular size of the arc connecting the compared positions), which is quite satisfactory for the problems considered in the book. A more detailed discussion of the methodology and some results of the analysis is

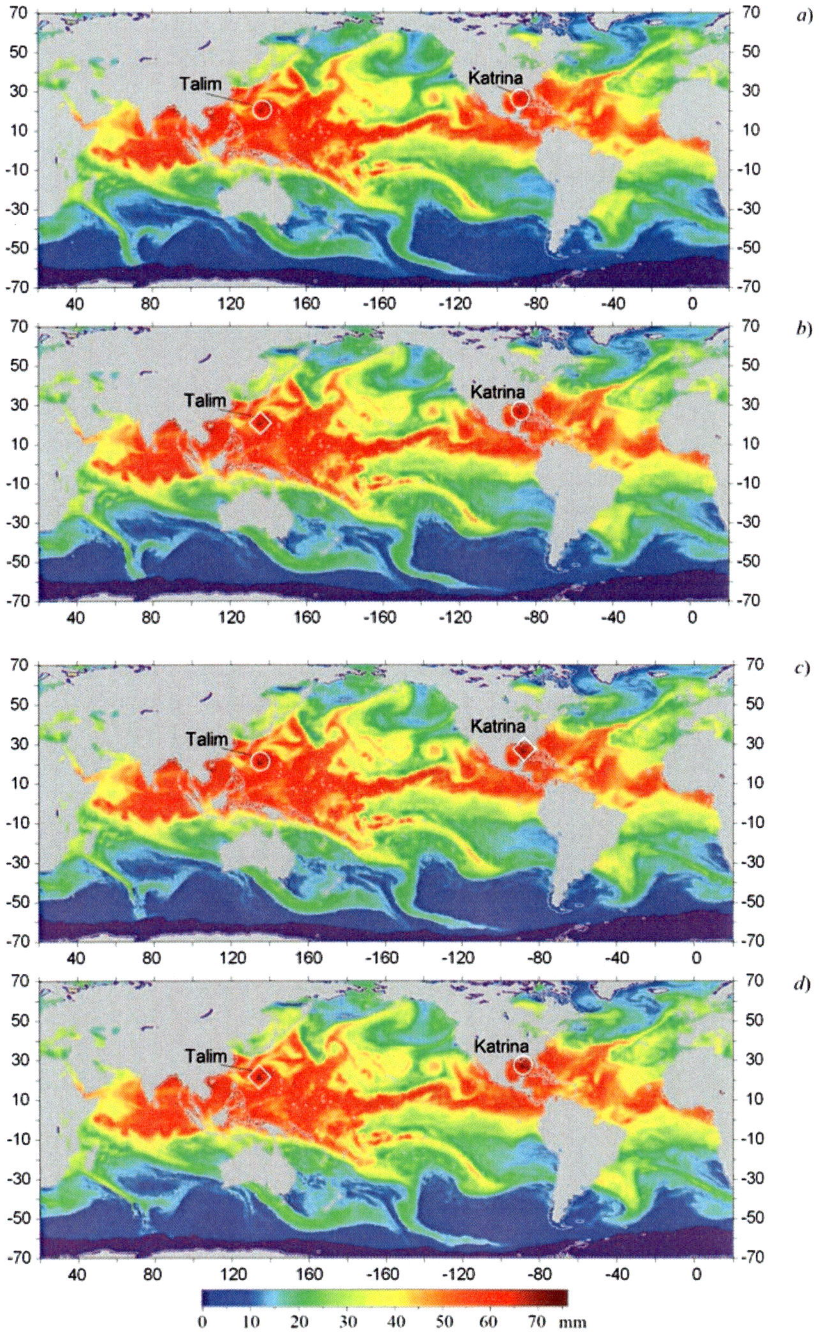

Fig. 3.10 The global TPW field on 08/28/2015 (color scale of TPW values below) at 12:00 UTC (**a**), 15:00 UTC (**b**), 18:00 UTC (**c**), 21:00 UTC (**d**) and the positions of the tropical cyclones Talim and Katrina (white circles and diamonds) at the corresponding time points

given in (Ermakov et al. 2011). Some aspects, capabilities and results of the analysis of the accuracy of the interpolation scheme are presented in (Ermakov et al. 2015b).

3.5 Iterative Extension of the Basic Scheme: A Multisensory Approach

The construction of reference radiothermal fields (or fields of geophysical parameters), as shown above, occupies an important place in satellite radiothermovision algorithms. The reference fields should provide complete coverage of the studied regions with measurements close by local time at all cells of the computational grid. This is achieved by combining data from several satellites close in observation time over the same areas (interval of about 1 h or less), followed by a smooth extrapolation of the obtained fields in the regions of data gaps (lacunae stapling).

Unfortunately, not in all cases this approach is directly applicable. So, for example, the local time of the ascending node of the satellite GCOM-W1 (instrument AMSR-2) is about 13:30. The difference with the local time of the ascending nodes of the DMSP *F*17 satellites (SSMIS instrument) and Coriolis (WindSat) for the entire time of AMSR-2 measurements was at least 4 h. In relation to the DMSP *F*16 satellite, this difference was about 4 h in August 2012, about 3.5 h in November 2013 and decreased to 3 h by March 2015. Thus, the direct combination of AMSR-2 data with SSMIS and WindSat data to construct reference fields was not possible. On the other hand, the use of exclusively AMSR-2 data in constructing reference fields gives too wide lacunae, and their "stapling" results in significant processing artifacts. Consequently, the joint application of all available data for constructing reference fields is difficult, and the simplest solution is to refuse to use AMSR-2 data, despite their unique spatial resolution.

At the same time, analysis of the realized spatiotemporal interpolation scheme revealed that "synchronization" of heterogeneous data series with its help is performed with an accuracy that is quite acceptable (Wimmers and Velden 2011; Observing Systems 2011) for a wide class of remote sensing tasks. In this regard, the possibility of an iterative algorithm for constructing reference fields arises, schematically depicted in Fig. 3.11 and illustrated by an example for the observation interval in November 2013.

On the left, in Fig. 3.11, the main steps of the algorithm are shown. At the first processing step, all available radiothermal data (products) for the considered time interval are collected. They are arranged in chronological order of the passage of satellites over the studied regions. In the case of solar-synchronous orbits, ordering is conveniently carried out according to local time of the ascending and descending nodes of the orbit. On the right in Fig. 3.11, a fragment of the time axis is shown, on which the arrows indicate the moments of the passage of the descending (down arrows) and ascending (up arrows) nodes of the DMSP *F*16, *F*17 and GCOM-W1

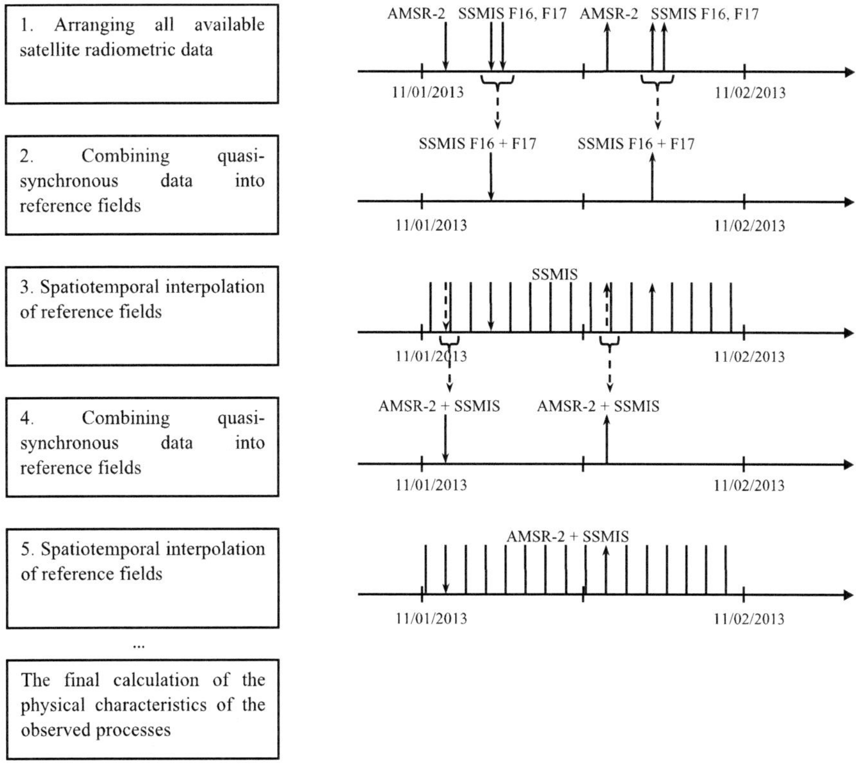

Fig. 3.11 An iterative scheme for constructing and spatiotemporal interpolating of the fields of geophysical parameters of global coverage according to satellite radiometric observations

satellites in November 2013. It is important that the time intervals between measurements by different instruments remain practically unchanged for all basins of the World Ocean over long observation intervals.

At the second step of the scheme, the data separated by small measurement intervals (about 1 h or less) are combined and the reference fields are constructed using the operation of lacunae stapling. In this example, only SSMIS data are subject to combining on a regular grid of 0.25°. AMSR-2 data is not involved in processing at this stage.

The third step implements the spatiotemporal interpolation of the obtained reference fields based on an algorithm for motion estimation and compensation. As a result, the fields of the geophysical parameter under study are formed with a time step of 1.5 h on a 0.25° grid. On the time axis they are shown by lines without arrows. Here, dashed lines show the source data of AMSR-2.

In the fourth step, the re-construction of reference fields occurs. This time, data not previously involved in the analysis are taken as a basis, which are supplemented in the areas of lacunae by fragments of interpolated fields obtained in the third step, which are "synchronized" with the first ones as much as possible. In this example,

AMSR-2 data are taken as the basis. They are resampled to a regular grid with a step of 0.125°; therefore, when supplementing them with SSMIS data built on a 0.25° grid, it makes sense to first bring the latter to the same, more detailed grid in order to preserve the detail of the main measurements (AMSR-2). Given the smoothness of the studied geophysical fields at the scales under consideration, for this purpose, it is proposed to perform a simple bilinear interpolation of the corresponding fragments of the SSMIS fields. It should be emphasized once again that although the main goal of the gap filling procedure is to minimize the solution perturbations associated with edge effects, the analysis of the interpolation accuracy showed a high quality of reconstructing the values of the geophysical parameter at all cells of the computational grid, including in the regions of lacunae (Ermakov et al. 2016).

The fifth step again implements the spatiotemporal interpolation of the obtained reference fields. The result, in the considered example, is a sequence of fields of a geophysical parameter built on a 0.125° grid with a time step of 1.5 h. The inclusion of AMSR-2 in the data processing in this case allows improving the spatial detail of the analyzed fields, which, in particular, contributes to a more accurate calculation of advection fields and estimates of the integral characteristics of processes (for example, latent heat fluxes) on a more detailed computational grid. It also opens the possibility of mutual calibration of data processing products according to different instruments due to their better "synchronization".

The steps of constructing reference fields and their spatiotemporal interpolation, in principle, can be carried out repeatedly, depending on the specific purpose of processing (for example, the inclusion of additional data series). Completing the processing, according to the general calculation scheme of satellite radiothermovision, is evaluating the integral physical characteristics (for example, advective latent heat fluxes when studying mesoscale and synoptic atmospheric processes) based on the constructed scalar fields of geophysical parameters and vector fields of advection (see Sect. 3.6).

As an example of constructing reference fields, the processing of the TPW fields W reconstructed according to the SSMIS ($F16$, $F17$ DMSP) and AMSR-2 (GCOM-W1) is illustrated below. Measurement data covered November 2013. For better detail, Fig. 3.12 shows only fragments of the constructed fields over the Indian Ocean (between 30° N and 30° S and 40° and 110° E) for November 1. Figure 3.12a contains a field (color scale in the lower right corner of the figure), reconstructed from SSMIS $F16$ measurements, supplemented by a small number of SSMIS $F17$ measurements to partially fill the lacunae (the difference in measurement times is about 1 h). As one can see, the lacunae are quite narrow and satisfactorily "stapled" by smooth extrapolation of the field, Fig. 3.12b. Figure 3.12c characterizes the change in the field during the time between measurements of AMSR-2 and SSMIS $F16$, visualizing the difference between the values of the field interpolated by the moment of AMSR-2 measuring and the reference field (Fig. 3.12b) in the fragment under consideration (the scale of differences is in the upper right corner of the figure).

Figure 3.12d contains a fragment of a field reconstructed from the AMSR-2 measurements for November 1. As noted above, SSMIS $F16$, $F17$ measurements are separated from AMSR-2 measurements by 3.5–4.5 h, which prevents the direct

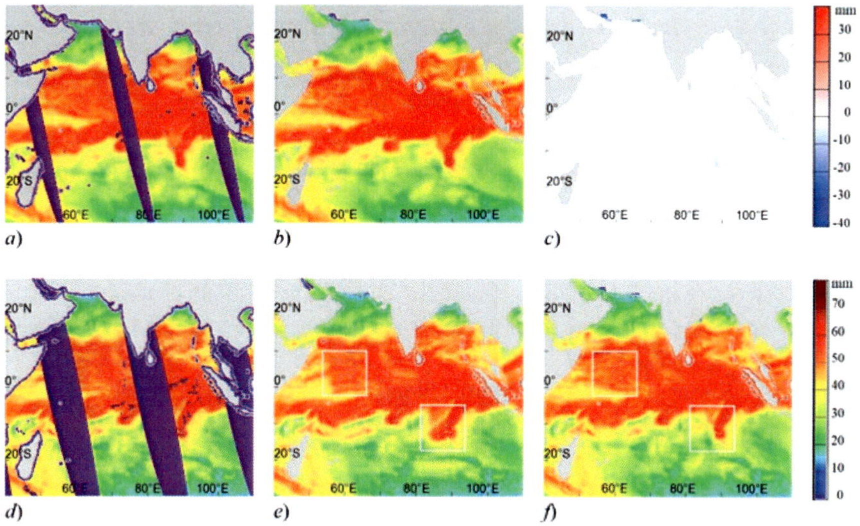

Fig. 3.12 The TPW field the atmosphere over the Indian Ocean on 11/01/2013, reconstructed according to SSMIS $F16$, $F17$ and AMSR-2 without and with interpolation (see comments in text)

integration of all these data to construct reference fields. On the other hand, in the case of AMSR-2, lacunae stapling due to their large width gives significant extrapolation errors, Fig. 3.12e. The optimal solution is to supplement the AMSR-2 data with the previously combined and interpolated SSMIS data, Fig. 3.12f. Some areas of the most noticeable differences in Figs. 3.12e, f are highlighted with white frames.

A more detailed discussion of the iterative multisensory algorithm for satellite radiothermovision together with additional processing examples is contained in (Ermakov et al. 2016). Using it, the evolutionary processes of a number of tropical cyclones were analyzed (including in the phases of rapid intensification); analysis results are included in Chap. 4.

3.6 New Opportunities in the Remote Sensing

Together, the considered algorithms of satellite radiothermovision provide a number of new possibilities for the quantitative description of dynamic processes. Before moving on to the main one, within the framework of the book, we should briefly dwell on some others.

3.6.1 *Joint Analysis of Independent Satellite Measurements*

The study of rapidly developing processes in the Earth's ocean-atmosphere system using observations from low-orbit satellites presents a methodological problem of "synchronization" of observation data. So, observation by each of the satellites of a given region of space from a sun-synchronous orbit is carried out only twice a day, and coverage of low latitudes is incomplete. As a result of this, objects of research that are rapidly passing through the phases of evolution can be registered only partially, or even fall out of the field of view of certain devices. The problem is partially solved by mechanically combining several series of observational data, which, however, can lead to significant artifacts (see Fig. 3.4). Satellite radiothermovision offers a more reasonable approach to combining independent and asynchronous observations to obtain consistent estimates of the local and integral parameters of the observed objects. This approach, in particular, is widely applied in the analysis of the evolution of tropical cyclones (Chap. 4), both for detailing the dynamics of the fields of total precipitable water and for their joint processing with fields of the ocean surface temperature (Ermakov et al. 2015b). The same approach is also effective in the problems of calibration and cross-calibration of satellite data and products of their processing (Ermakov et al. 2016).

It should also be noted that the implemented spatiotemporal interpolation scheme allows to switch from analyzing data in the "time base" format (with constant local time of observations) to analyzing "instant" global geophysical fields (with constant universal time). Indeed, the construction of such "instantaneous fields" amounts to a regrouping of elements of arrays of interpolated fields: filling each node of the computational grid with an interpolated value at local time corresponding to (as close as possible) a certain fixed point in time, ensures the formation of such an "instant field" (Ermakov et al. 2016).

3.6.2 *Ensuring Spatial Connectivity of Fragmented Observations*

The satellite radiothermovision approach provides the calculation of the fields of atmospheric geophysical parameters on a global geographic grid without data gaps (lacunae), due to the use of information redundancy of multiple periodic radiothermal observations from different satellites. The completeness of coverage with data from the observed basins of the oceans is extremely important when studying the dynamics of extended atmospheric processes, the horizontal dimensions of which exceed (sometimes–substantially) the swath width of a satellite instrument. Satellite radiothermovision solves the problem of fragmentation of such objects of research and removes a number of significant methodological difficulties associated with the automation of processing large arrays of measurements including the calculation of their integral characteristics. This advantage of the developed approach is clearly manifested, for example, in the context of the study of atmospheric rivers, the results of which are presented in Chap. 5.

3.6.3 The Study of Vector Fields of Advection

One of the steps of the spatiotemporal interpolation scheme of geophysical atmospheric fields is the calculation of the vector field of horizontal displacements (advection), which is the solution to the problem of "motion estimation". Since the reference fields used in the calculations have a clear coordinate-time reference, the reconstructed vector motion fields as a result of normalization and geometric correction can be recalculated into the fields of average (in the interval between successive observations) advection velocities. Advection velocity fields can be used by themselves to analyze regional and global dynamics of air masses and study the structure of atmospheric circulation. In addition, together with the reconstructed scalar fields of the corresponding geophysical parameters, they can be used to calculate the integral characteristics of the observed processes, such as the power of latent heat fluxes through a given boundary (Ermakov et al. 2017, 2019b). This feature is discussed in more detail below. It underlies the analysis of the evolution of tropical cyclones (Chap. 4), as well as the study of a number of important characteristics of atmospheric rivers (Chap. 5) and the structure of global atmospheric circulation (Chap. 6).

3.6.4 Calculation of the Integral Characteristics of Mass and Energy Transfer

As noted above, the simultaneous restoration of the scalar field of a geophysical parameter and the vector field of its advection (M-field) allows the calculation of the parameter flow through arbitrarily given boundaries. If it corresponds to the total precipitable water of the atmosphere, then its flow (up to normalization) gives the total power of the latent heat flux (integral over the height of the atmosphere) through the surface with vertical side walls and the projection onto the Earth's surface in the form of a given boundary (contour).

Let us first consider the scheme for calculating the flow through some circuit element (Fig. 3.13). To study the evolution of tropical cyclones (Chap. 4), circular contours with given centers and radii are convenient, therefore, along with a rectangular coordinate system of nodes of the computational grid, we introduce a radial coordinate system with an origin at the proposed center of the contour (x_0, y_0). Let the contour element $\vec{l} = \left(l_x, l_y\right)$ under consideration pass through the end of the radius-vector $\vec{r}$ with the coordinates (x, y) that make up the angle α with the horizontal axis (Fig. 3.12), so that $x = x_0 \cos\alpha,\ y = y_0 \sin\alpha$. Denote the external normal (in the

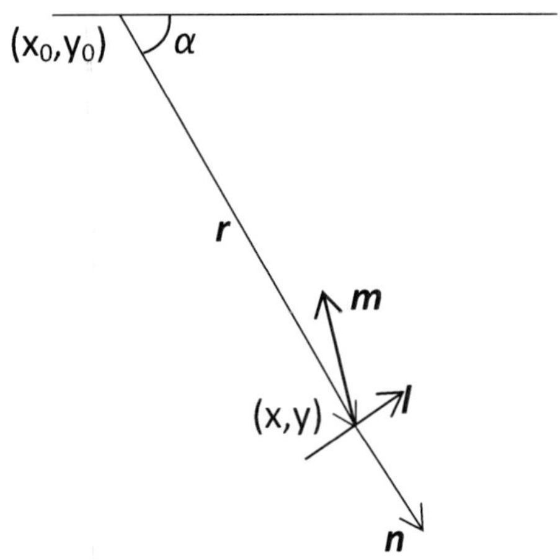

Fig. 3.13 Scheme for calculating the flux through the contour element

direction $\vec{r}$) to the contour element at the point (x, y) as $\vec{n} = (n_x, n_y) = (\cos\alpha, \sin\alpha)$ and the value of the M-field at the same point as $\vec{m} = (m_x, m_y)$.

To calculate the flux through the contour element in standard units, it is necessary to transit from the vector $\vec{m}$ in the coordinates (indices) of the nodes of the calculated image grid to the velocity vector $\vec{v}$ in m/s relative to the Earth and take into account projective distortions. Let the grid lines be numbered so that the coordinate $y = 0$ corresponds to 90° N and increases to the south (from top to bottom), and the coordinate $x = 0$ corresponds to 180° W and increases to the east (from left to right). Changing any index by one leads to a change in the angular coordinate (latitude or longitude) by an amount $s = 0.2°$ equal to the grid step. Therefore, the coordinates (x, y) are expressed in terms of longitude λ and latitude θ as follows:

$$\lambda = s \cdot x - \pi, \theta = \frac{\pi}{2} - s \cdot y, \tag{3.33}$$

and the angular measures of the vector $\vec{m} = (m_x, m_y)$ in the coordinates of longitude λ and latitude θ are:

$$m_\lambda = s \cdot m_s; m_\theta = -s \cdot m_y. \tag{3.34}$$

Elementary relations of spherical geometry give:

$$\vec{v} = \frac{Rs}{\Delta t}(m_x \cdot \sin(sy), -m_y), \tag{3.35}$$

where $R = 6371$ km is the average radius of the Earth, Δt is the time step of interpolation.

Similarly, the vector of the unit normal to the contour element after correction of projective distortions is equal to:

$$\overrightarrow{n'} = \frac{(\sin(sy)\cos\alpha, -\sin\alpha)}{\sqrt{\sin^2(sy)\cos^2\alpha + \sin^2\alpha}}, \tag{3.36}$$

and the length of the contour element on the surface of the Earth:

$$\left|\overrightarrow{l'}\right| = R \cdot s \cdot r \cdot d\alpha \cdot \sqrt{\sin^2(sy)\sin^2\alpha + \cos^2\alpha}, \tag{3.37}$$

where r is the length of the vector $\vec{r}$ (Fig. 3.12) in the coordinates of the satellite image (pixels), and $\mathrm{d}\alpha$ is the angle at which the contour element $\vec{l}$ is visible from the grid cell (x_0, y_0).

Then the flow of the velocity field $\vec{v}$ through the contour element with the normal $\overrightarrow{n'}$ at the point (x, y) is:

$$\mathrm{d}\Phi(x, y) = -\left(\vec{v} \cdot \overrightarrow{n'}\right) \cdot \left|\overrightarrow{l'}\right|$$
$$= -\frac{R^2 s^2 r\left(m_x \sin^2(sy)\cos\alpha + m_y \sin\alpha\right)\sqrt{\sin^2(sy)\sin^2\alpha + \cos^2\alpha}}{\Delta t\sqrt{\sin^2(sy)\cos^2\alpha + \sin^2\alpha}} d\alpha. \tag{3.38}$$

The sign is selected so that the positive flux is directed inside the contour.

From here, it is easy to transit to the latent heat flux through the same contour element:

$$\mathrm{d}Q(x, y) = q \cdot W(x, y) \cdot d\Phi(x, y), \tag{3.39}$$

where $q = 2260\,\mathrm{KJ/kg}$ is the specific heat of vaporization, $W(x, y)$ is the total precipitable water of the atmosphere at a point (x, y) in kg/m^2 (or in mm).

The numerical integration of expression (3.39) over an angle α from 0 to 2π gives the value of the latent heat flux through a circular contour of radius r (in pixels) centered at the cell (x_0, y_0) of the computational grid. In view of (3.38), the final expression (3.39) is closed with respect to the input data, i.e. the calculation can be carried out directly based on the results of the previous spatiotemporal interpolation (synchronous pair of W field and M-field). For a contour of radius r, it is reasonable to choose the integration step $\mathrm{d}\alpha$ from the relation $r \cdot \mathrm{d}\alpha \sim 1$, which corresponds to the length of the contour elements, approximately equal to the cell size of the coordinate grid.

It should be noted that the expressions for fluxes (3.38) and (3.39) were obtained for a fixed contour. Sometimes it is preferable to take into account the "drift" of the contour relative to the Earth. Then in the expression (3.38) it is necessary to substitute the velocity $\overrightarrow{v'} = \vec{v} - \vec{u}$ instead of the velocity $\vec{v}$, where $\vec{u}$ is the drift velocity. The choice of drift velocity is determined by the statement of the problem. In some cases, as a first approximation, one can use the average value of $\vec{v}$ over the contour:

$$\vec{u} = \frac{1}{N}\sum_{i=1}^{N}\vec{v_i}, \tag{3.40}$$

where $\vec{v_i}$ is the velocity value calculated according to (3.30) for the ith grid cell included in the contour, N is the total number of cells included in the contour.

In (Ermakov et al. 2017), the derivation of similar relations for an arbitrarily given boundary and a special case for the meridional component of the latent heat flux are presented. These results were used to study atmospheric rivers (Chap. 5) and the characteristics of global atmospheric circulation (Chap. 6).

3.7 Concluding Remarks to this Chapter

1. The research results presented in this Chapter relate to the physical, mathematical and methodological foundations of the satellite radiothermovision approach.
2. The chapter fully describes the basis of a computational scheme which allows robust spatiotemporal interpolating of both global radiothermal fields measured from solar-synchronous satellites and fields of geophysical parameters of the ocean-atmosphere system retrieved from these remote data.
3. The described approach provides highly detailed (with a spatial step of up to 0.125 geographical degrees and a time step of 1.5 h, sometimes better) retrieval of the state of the Earth's atmosphere without data gaps.
4. The analysis proved acceptable precision of the interpolated fields which make it possible to use them in studying the dynamics and energy aspects of mesoscale and synoptic processes. The most important feature of the proposed approach is the ability to calculate the atmospheric latent heat flux from remote data. The next three chapters will treat the issue of a practical implementation of the satellite radiothermovision approach in detail.

References

Akvilonova AB, Kutuza BG (1978) Radioteplovoe izluchenie oblakov (Microwave radiothermal emission of clouds) Radiotekhnika i elektronika 23(9):1792–1806 (in Russian)

Alexanin AI, Alexanina MG, Karnatsky AY (2011) Avtomaticheskiy raschet skorostey peremescheniy ledovyh poley (Automatic calculation of movement velocities of sea ice fields). Sovremennye problemy distantsionnogo zondirovaniya Zemli iz kosmosa 8(2):9–17 (in Russian)

Alexanin AI, Alexanina MG, Karnatsky AY (2013) Avtomaticheskiy raschet skorostey poverhnostnyh techeniy okeana po posledovatelnosti sputnikovyh izobrazheniy (Automatic computation of sea surface velocities on a sequence of satellite images). Sovremennye problemy distantsionnogo zondirovaniya Zemli iz kosmosa 10(2):131–142 (in Russian)

Anandan P (1989) A computational framework and an algorithm for the measurement of visual motion. Int J Comput Vis 2(3):283–310

Armand NA, Polyakov VM (2005) Radio propagation and remote sensing of the environment. CRC Press, Boca Raton, 448p

AviSynth (2002) A powerful nonlinear scripting language for video. AviSynth. http://avisynth.org/

Barron JL, Fleet DJ, Beauchemin SS (1994) Performance of optical flow techniques. Int J Comput Vis 12(1):43–77

Basharinov AE, Gurvich AS, Egorov ST (1974) Radioizlucheniye Zemli kak planet (Radioemission of the Earth as a planet). Nauka, Moscow, 187p (in Russian)

Bindschadler RA, Scambos TA (1991) Satellite-image-derived velocity field of an Antarctic ice stream. Science 252(5003):242–246

Bobkov VA, Kazansky AV, Morozov MA, Shchebenkova AA (2003) Relaksatsionno-konturnyy algoritm opredeleniya vektorov morskih techeniy po sputnikovym izobrazheniyam i ego sinopticheskaya verifikatsiya (The relaxation-contour algorithm for determining the vectors of sea currents from satellite images and its synoptic verification). Avtometriya (1):73–81 (in Russian)

Bowen MM, Emery WJ, Wilkin JL, Tildesley PC, Barton IJ, Knewtson R (2002) Extracting multi-year surface currents from sequential thermal imagery using the maximum cross-correlation technique. J Atmos Ocean Technol 19(10):1665–1676

Breaker LC, Krasnopolsky VM, Rao DB, Yan X-H (1994) The feasibility of estimating ocean surface currents on an operational basis using satellite feature tracking method. Bull Am Meteorol Soc 75(11):2085–2095

Dworak R, Key JR (2009) Twenty years of polar winds from AVHRR: validation and comparison with ERA-40. J Appl Meteorol Climatol 48(1):24–40

Eigenwillig N, Fischer H (1982) Determination of midtropospheric wind vectors by tracking pure water vapor structure in METEOSAT water vapor image sequences. Bull Am Meteorol Soc 63(1):44–57

Emery WJ, Thomas AC, Collins MJ, Crawford WR, Mackas DL (1986) An objective method for computing advective surface velocities from sequential infrared satellite images. J Geophys Res 91(C11):12865–12878

Ermakov DM (2017) "Interlon": Programmnoye obespecheniye dlya prostranstvennoy interpolyatsii sutochnyh poley productov sputnikovogo radioteplovogo monitoringa Zemli (Software for combining data processing products of the daily satellite radiothermal monitoring of the Earth). Svidetel'stvo gosudarstvennoy registratsii programmy dlya EVM # 2017663104. (Certificate of state registration of computer programs No. 2017663104.) Registration date 11/24/2017 (in Russian)

Ermakov DM (2018) Global circulation of latent heat in the Earth's atmosphere according to data from satellite radiothermovision. Izv Atmos Ocean Phys 54(9):1223–1243

Ermakov D, Chernushich A, Sharkov E, Shramkov Y (2011) Stream handler system: an experience of application to investigation of global tropical cyclogenesis (electronic resource). In: Proceedings of 34th international symposium on remote sensing of environment, Sydney, 10–15 April, 2011. URL: http://www.isprs.org/proceedings/2011/ISRSE-34/211104015Final00456.pdf

Ermakov DM, Sharkov EA, Chernushich AP (2015a) Assessment of the accuracy of the interpolation scheme of satellite radiothermovision. Sovremennye Problemy Distantsionnogo Zondirovaniya Zemli iz Kosmosa (Curr Probl Remote Sens Earth Space) 12.(2):77–88 (in Russian)

Ermakov DM, Sharkov EA, Chernushich AP (2015b) Satellite radiothermovision of atmospheric mesoscale processes: case study of tropical cyclones. Int Arch Photogramm Remote Sens Spat Inf Sci ISPRS Arch XL(7/W3):79–186

Ermakov DM, Sharkov EA, Chernushich AP (2016) A multisensory algorithm of satellite radiothermovision. Izv Atmos Ocean Phys 52(9):1172–1180

Ermakov DM, Sharkov EA, Chernushich AP (2017) Satellite radiothermovision on synoptic and climatically significant scales. Izv Atmos Ocean Phys 53(9):973–978

Ermakov DM, Raev MD, Chernushich AP, Sharkov EA (2019a) Algorithm for construction of global ocean-atmosphere radiothermal fields with high spatiotemporal sampling based on satellite microwave measurements. Izv Atmos Ocean Phys 55(9):1041–1052. https://doi.org/10.1134/s0001433819090159

Ermakov DM, Sharkov EA, Chernushich AP (2019b) Evaluation of tropospheric latent heat advective fluxes over the ocean by the animated analysis of satellite radiothermal remote data. Izv Atmos Ocean Phys 55(9):1125–1132. https://doi.org/10.1134/s0001433819090160

Ermakov DM, Smirnov MT (2001) Problems of superposition of satellite optical and microwave sensors data for their complex analysis. Issledovanie Zemli iz Kosmosa (Earth Res Space) (2):45–55 (in Russian)

FileX (2000) Intelligent vision systems. FileX. https://www.thefilex.com/software

Fleet DJ, Weiss Y (2005) Optical flow estimation. In: Paragios N, Chen Y, Faugeras O (eds) Chapter 15: Mathematical models in computer vision: The handbook. Springer. Berlin, pp 239–258

Gaikovich KP (2003) High-sensitivity radiometry of air-water interface fast temperature and heat flux variances. IEEE Trans Geosci Remote Sens 41(12):2760–2766

Gaikovich K, Troitsky RV (1998) Dynamics of temperature profile, heat, and mass exchange through air-water interface by measurements of thermal radio emission evolution at 60 GHz. IEEE Trans Geosci Remote Sens 36(1):341–344

Garbe CS, Ommer B (2013) Parameter estimation in image processing and computer vision. In: Bock HG et al (eds) Model based parameter estimation: Theory and applications. Springer, Berlin, pp 311–334

Gledzer EB, Golitsyn GS (2010) Skeyling i konechnye ansambli chastits v dvizhenii s pritokom energii (Scaling and finite ensembles of particles in motion with an influx of energy). Doklady Akademii Nauk 443(4):446–470 (in Russian)

Gorbunov ME, Kutuza BG (2018) Cosmos-243 as the starting point for the development of microwave radiometry methods of the Earth's atmosphere and surface. Izv Atmos Ocean Phys 54(3):275–281

Grankov AG, Milshin AA (2016) Microwave radiation of the ocean-atmosphere: boundary heat and dynamic interaction. 2nd edn. Springer, Berlin, 193p

Grishin SV, Vatolin DS, Lukin AS, Putilin SY, Strelnikov KN (2008) Obzor blochnyh metodov otsenki dvizheniya v tsifrovyh video signalah (An overview of block methods for estimating motion in digital video signals). In: Programmnye sredstva i instrumenty: tematicheskiy sbornik N 9 (Software systems and tools: thematic collection No 9). MGU, Moscow, pp 50–62 (in Russian)

Grzywacz NM, Yuille AL (1990) A model for the estimation of local image velocity by cells in the visual cortex. Proc R Soc Lond B239(1295):129–161

Gurvich AS, Kutuza BG (2010) "Kosmos-243" – pervyi v mire eksperiment po issledovaniyu Zemli iz kosmosa radiofizicheskimi metodami ("Kosmos-243"— the first experiment in the World on the Earth research from space using radiophysical methods). Issledovanie Zemli iz kosmosa (2):14–25 (in Russian)

Hartman DL (1999) Radiation, clouds, water vapor, precipitation, and atmospheric circulation. In: King MD (ed) EOS science plan: the state of science in the EOS program. NASA, pp 39–114

Haug T, Kääb A, Skvarca P (2010) Monitoring ice shelf velocities from repeat MODIS and landsat data—a method study on the larsen C ice shelf, antarctic peninsula, and 10 other ice shelves around Antarctica. Cryosphere 4(2):161–178

Horn BKP, Schunck BG (1981) Determining optical flow. Artif Intell 17(17):185–203

Kääb A (2002) Monitoring high-mountain terrain deformation from repeated air- and spaceborne optical data: examples using digital aerial imagery and ASTER data. ISPRS J Photogramm Remote Sens 57(1–2):39–52

Kääb A, Vollmer M (2000) Surface geometry, thickness changes and flow fields on creeping mountain permafrost: automatic extraction by digital image analysis. Permafr Periglac Process 11(4):315–326

Kadygrov EN, Shur GN, Viazankin AS (2003) Investigation of atmospheric boundary layer temperature, turbulence, and wind parameters on the basis of passive microwave remote sensing. Radio Sci 38(3):8048, 13.1–13.12

Key JR, Santek D, Velden CS, Bormann N, Thepaut J-N, Riishojgaard LP, Zhu Y, Menzel WP (2003) Cloud-drift and water vapor winds in the polar regions from MODISIR. IEEE Trans Geosci Remote Sens 41(2):482–492

Kutuza BG, Danilychev MV, Yakovlev OI (2016) Sputnikovyy monitoring Zemli: Mikrovolnovaya radiometriya atmosfery i poverhnosti (Satellite monitoring of the Earth: Microwave radiometry of the atmosphere and surface). LENAND, Moscow, 336p (in Russian)

Kutuza BG et al (2019) Pervyy v mire eksperiment po mikrovolnovomu zondirovaniyu Zemli iz kosmosa na sputnike "Kosmos-243" (The world's first experiment on microwave Earth sensing from space on the "Kosmos-243" satellite). Sovremennye Problemy Distantsionnogo Zondirovaniya Zemli iz Kosmosa 16(6):9–30 (in Russian)

Lazzara MA, Dworak R, Santek DA, Hoover BT, Velden CS, Key JR (2014) High-latitude atmospheric motion vectors from composite satellite data. J Appl Meteorol Climatol 53(2):534–547

Little JJ, Bulthoff HH, Poggio TA (1988) Parallel optical flow using local voting. In: Proceedings of the 2nd IEEE international conference on computer vision. Tampa, pp 454–459

Little JJ, Verri A (1989) Analysis of differential and matching methods for optical flow. In: IEEE workshop on visual motion. Irvine, pp 173–180

Liu WT (1986) Statistical relation between monthly mean precipitable water and surface-level humidity over global oceans. Mon Weather Rev 114(8):1591–1602

Liu WT (1988) Moisture and latent heat flux variabilities in the Tropical Pacific derived from satellite data. J Geophys Res 93(C6):6749–6760

Liu W, Ribeiro E (2011) A survey on image-based continuum-body motion estimation. Image Vis Comput 29(8):509–523

Liu WT, Tang W (2005) Estimating moisture transport over oceans using space-based observations. J Geophys Res 110(D10):D10101. https://doi.org/10.1029/2004jd005300

Lucchitta BK, Ferguson HM (1986) Antarctica: measuring glacier velocity from satellite images. Science 234(4780):1105–1108

Menzel WP (2001) Cloud tracking with satellite imagery: from the pioneering work of Ted Fujita to the present. Bull Am Meteorol Soc 82(1):33–47

Nagel H-H (1987) On the estimation of optical flow: relations between different approaches and some new results. Artif Intell 33(3):299–324

Nagel H-H, Enkelmann W (1986) An investigation of smoothness constraints for the estimation of displacement vector fields from image sequences. IEEE Trans Pattern Anal Mach Intell 8(5):565–593

Nam KM, Kim J-S, Park R-H, Shim YS (1995) A fast hierarchical motion vector estimation algorithm using mean pyramid. IEEE Trans Circuits Syst Video Technol 5(4):344–351

Nerushev AF, Kramchaninova EK (2011) Method for determining atmospheric motion characteristics using measurements on geostationary meteorological satellites. Izv Atmos Ocean Phys 47(9):1104–1113

Nieman SJ, Menzel WP, Hayden CM, Gray D, Wanzong ST, Velden CS, Daniels J (1997) Fully automated cloud-drift winds in NESDIS operations. Bull Am Meteorol Soc 78(6):1121–1133

Observing Systems (2011) Observing systems capacity analysis and review tool. World Meteorological Organization. URL: https://www.wmo-sat.info/oscar/

Obukhov AM, Basharinov AE, Vasiliev YV, Gurvich AS, Kutuza BG, Mitnik LM (1971) Issledovaniye atmosfery po sobstvennomu radioteplovomu izlucheniyu na ISZ "Kosmos-243" (Investigation of the atmosphere using its own thermal radiation at the cosmos-243 satellite). Kosmicheskiye issledovaniya 9(1):66–73 (in Russian)

Petty GW (1990) On the response of the special sensor microwave/imager to the marine environment—implications for atmospheric parameter retrievals. A dissertation … for the degree of doctor of philosophy. University of Washington, Seattle, 313p

Pokrovskaya IV, Sharkov EA (2006) Tropical cyclones and tropical disturbances of the world ocean: chronology and evolution. Version 3.1 (1983–2005). M., Poligraph servis, 728p

Richardson IEG (2003) H.264 and MPEG-4 video compression. Wiley, Chichester, 306p

Robertson FR, Bosilovich MG, Roberts JB, Reichle RH, Adler R, Ricciardully L, Berg W, Huffman GJ (2014) Consistency of estimated global water cycle variations over the satellite era. J Clim 27(16):6135–6154

Ruzmaikin A (2014) Klimat kak igra sluchaya (Climate as a game of chance). Uspekhi Fizicheskikh Nauk 184(3):297–311 (in Russian)

Savorskiy VP (1992) Prostranstvenno-vremennaya structura SVCh radioteplovogo polya sistemy ocean-atmosfera (Spatiotemporal structure of the microwave radiothermal field of the ocean-atmosphere system). Dis. … Dr. Phys. Sciences. Moscow, 1992p (in Russian)

Scambos TA, Dutkiewicz MJ, Wilson JC, Bindschadler RA (1992) Application of image cross-correlation to the measurement of glacier velocity using satellite image data. Remote Sens Environ 42(3):177–186

Sharkov EA (2003) Passive microwave remote sensing of the Earth: Physical foundations. Springer, Chichester

Sharkov EA, Kuzmin AV, Vedenkin NN, Jeong S, Ermakov DM, Kvitka VE, Kozlova TO, Komarova NY, Minaev PY, Park IlH, Pashinov EV, Pozanenko AS, Prasolov VO, Sadovskii IN, Sazonov DS, Sterlyadkin VV, Khapin YB, Hong G, Chernenko AM (2019) Convergence space experiment: scientific objectives, onboard equipment, and methods of solving reverse problems. Izv Atmos Ocean Phys 55(9):1437–1456. https://doi.org/10.1134/s0001433819090469

Svejkovsky S (1988) Sea surface flow estimation from advanced very high resolution radiometer and costal zone color scanner satellite imagery: a verification study. J Geophys Res 93(C6):6735–6743

Vastano AC, Borders SE (1984) Sea surface motion over an anticyclonic eddy on the Oyashio front. Remote Sens Environ 16(1):87–90

Velden CS, Hayden CM, Nieman SJ, Menzel WP, Wanzong S, Goerss JS (1997) Upper-tropospheric winds derived from geostationary satellite water vapor observations. Bull Am Meteorol Soc 78(2):173–195

Velden CS, Daniels J, Stettner D, Santek D, Key J, Dunion J, Holmlund K, Dengel G, Bresky W, Menzel P (2005) Recent innovations in deriving tropospheric winds from meteorological satellites. Bull Am Meteorol Soc 86(2):205–223

Vogel C, Bauder A, Schindler K (2012) Optical flow for glacier motion estimation. ISPRS Ann Photogramm Remote Sens Spat Inf Sci I-3:359–364

Weickert J, Schnörr C (2001) A theoretical framework for convex regularizers in PDE-based computation of image motion. Int J Comput Vis 45(3):245–264

Weng F, Zou X, Wang X, Yang S, Goldberg MD (2012) Introduction to Suomi national polar-orbiting partnership advanced technology microwave sounder for numerical weather prediction and tropical cyclone applications. J Geophys Res 117(D19): D19112. https://doi.org/10.1029/2012jd018144

Wentz F (1997) A well-calibrated ocean algorithm for special sensor microwave/imager. J Geophys Res 102(C4):8703–8718

Wimmers AJ, Velden CS (2011) Seamless advective blending of total precipitable water retrievals from polar orbiting satellites. J Appl Meteorol Climatol 50(5):1024–1036

Yamaguchi Y, Tanaka S, Odajima T, Kamai T, Tsuchida S (2003) Detection of a landslide movement as geometric misregistration in image matching of SPOT HRV data of two different dates. Int J Remote Sens 24(18):3523–3534

Zhu Y, Newell RE (1994) Atmospheric rivers and bombs. Geophys Res Lett 21(18):1999–2002

Chapter 4
Satellite Radiothermovision of Tropical Cyclones

This chapter demonstrates the unique capabilities of the satellite radiothermovision approach for studying the processes of evolution of tropical cyclones (TC). A general description of these atmospheric phenomena and the justification of the importance of their study by remote sensing methods is given in Chap. 2. Satellite radiothermovision provides the possibility of calculating convergent and divergent latent heat fluxes generated in the vicinity of an existing TC, and, as shown in the examples of processing specific data of satellite radiometry observations, which play the determining role in his energy budget. To this end, the first section of the chapter indicates the most common characteristics and methods for assessing the energy (power) of a TC and briefly discusses the main factors that can influence its intensification and dissipation, and some approaches to their analysis. The second section describes the data used and the methods of their processing in the framework of the satellite radiothermovision approach (Ermakov et al. 2019a, b, c). The third section describes the results of an analysis of the evolution of a number of specific TCs in the field of total precipitable water (TPW) of the atmosphere, and shows the relationship between the intensification (dissipation) of TCs and the formation of convergent (divergent) latent heat fluxes that are sufficient in absolute value to explain the evolution of TCs (Ermakov et al. 2019b, 2015). The fourth section provides examples of the integrated processing of combined satellite remote sensing data that allow the study of the evolution of TCs simultaneously in the fields of TPW and the temperature of the surface layer of the ocean (SST) (Ermakov et al. 2015). Finally, in the fifth section, using the analysis of the evolution of a system of interacting TCs (twin typhoons) as an example, the flexibility is demonstrated of adapting the previously used research methods to study a wide range of diverse mesoscale atmospheric processes (Ermakov et al. 2017).

D. M. Ermakov, *Satellite Radiothermovision of Atmospheric Processes*, Springer Praxis Books,
https://doi.org/10.1007/978-3-030-57085-9_4

4.1 Energy Characteristics and Energy Balance of a Tropical Cyclone

The section consists of two parts. The first indicates the most common energy characteristics of the TC. The second briefly discusses the main factors that can influence its intensification and dissipation, and some approaches to the analysis of these factors.

4.1.1 Tropical Cyclone Energy Characteristics

Historically, as the main parameter characterizing the energy of a tropical cyclone, its "intensity" is used, equal to the maximum sustainable horizontal wind speed V_{max} (m/s) in the eye wall of the TC (Palmén and Newton 1969). Strictly speaking, even based on the V_{max} dimension, it cannot be considered the energy characteristic of the TC. However, taking into account the similarity of the spatial structure of all TCs and the common mechanisms of their evolution, it is obvious that higher V_{max} values correspond to TCs with a higher kinetic energy. Variants of estimates of the kinetic energy of TCs based on the known value of their intensity may, however, somewhat vary (Emanuel 2005; Golitsyn 2008). The undoubted advantage of the characteristic is the possibility of its remote determination based on satellite data (Dworak 1975). It should also be noted that a close (negative) correlation with V_{max} demonstrates the value of the minimum pressure in the center of the eye of the TC, P_{min}.

The question of the "power" of the TC, i.e. the amount of energy generated by him per unit time causes serious discussion, including because of terminological differences. C. Landsea (Landsea 2014) indicates two methods for estimating the power of a TC. The first is the generation of energy released as a result of the condensation of atmospheric water vapor. For this case, he gives an average estimate of the order of $0.6 \cdot 10^{15}$ W, or 0.6 PW. The second method of evaluation belongs to K. Emanuel (Emanuel 1999), who strongly objects to the above value. He considers as the power of the TC the kinetic energy generation rate, which for a mature TC in a quasi-stationary state is estimated from the dissipative losses associated with friction, and reaches a value of the order of $1.5 \cdot 10^{12}$ W or 0.0015 PW. It should be noted that, taking into account the low efficiency of the TC for converting thermal energy into kinetic energy, which, according to estimates (Palmén and Newton 1969) does not exceed units of percent, both values are in good agreement with each other. The principal issue, therefore, is the source of heat input to the TC in volumes characterized by the first of the estimates and sufficient to maintain the corresponding kinetic energy of the TC. It should also be emphasized that the estimate of K. Emanuel should be substantially (possibly several times) increased in the analysis of the unsteady process of rapid intensification of TCs, when the kinetic energy generation rate exceeds dissipative losses so that the TC intensity V_{max} increases by 15 m/s or more per day.

4.1.2 Tropical Cyclone Energy Balance Factors

As it is known, the ocean is the main accumulator of solar radiation incident on the Earth. Therefore, when studying the processes of evolution (especially, rapid intensification) of TCs, it is important to study the role of the ocean as the dominant source of stored energy. This role is twofold, and, first of all, the "convective" hypothesis deserves attention, which proposes a mechanism for direct (vertical) energy transfer in the form of explicit and latent heat from the ocean to an existing TC to explain the evolution of the TC. In support of this hypothesis, both some observational data and simulation results are advanced, showing that heat transfer between the atmosphere and the active layer of the ocean can be intensified many times during hurricane winds realized in a mature TC. Nevertheless, as will be shown below, even extreme estimates of the rate of this heat transfer are unsatisfactory for explaining the full power of a mature TC, which is the main defect of the convective hypothesis.

Of the observation data mentioned above, the so-called "cold wake" (or "cold trace") should be noted first of all (Ivanov and Pudov 1977; Korolev et al. 1990; Hart et al. 2007; Dare and McBride 2011) which is the negative SST anomaly often observed in the ocean after the TC passing above it. According to some researchers, the "cold wake" is an indicator of active heat transfer from the ocean to the atmosphere (Lin et al. 2014). It should be noted, however, that when analyzing large ensembles of data, the correlations between the time course of SST anomalies and characteristics V_{max} and P_{min} of TCs are low (Gentemann et al. 2000), which does not allow to reliably reveal the patterns of their mutual changes. In (Zhao et al. 2008), an example is given of two typhoons that formed a "cold wake" in the South China Sea before going to land. The first typhoon was more intense, but moved faster to the shore. The second, weaker, for some time remained almost in one place above the ocean. Both of them created SST anomalies comparable in area and size (ibid., Figures 2,3), but the intensity of the second did not increase. On the contrary, it continued to decrease evenly until the final filling of the TC after going to land.

It should be taken into account that, in addition to heat exchange processes, an important role in the change in SST is played by the rise of deep waters under the influence of a hurricane, causing mixing of the upper layer of the ocean and the generation of an ocean vortex. Thus, the formation of SST anomalies to a significant and possibly decisive measure depends on the temperature of deep waters (and the depth of the warmed layer of the ocean). M.S. Permyakov points out (Permyakov 2007) based on statistics from ship observations that "a trace of a typhoon in the ocean can be warm". As follows from recent modeling results (Kudryavtsev et al. 2019a, b), the formation and sign of the SST anomaly is largely associated with the vertical temperature profile of the ocean.

It must be emphasized that the mechanisms of the origin of TCs (i.e., the formation of their mature forms from the initial tropical depressions) and the mechanisms of intensification of mature TCs can differ significantly (Palmén and Newton 1969; Frank 1977; Gray 1982; Rotunno and Emanuel 1987; Montgomery and Farrell 1993; Melnikov and Smulsky 1997; Permyakov 2007; Golitsyn 2008; Sharkov 2012; Fritz and Wang 2014; Levina 2018). In the early stages, when a stable vortex is not formed, the processes of vertical heat transfer must have a defining character, although, as noted in (Palmén and Newton 1969, p. 505), "it is generally agreed that a "convective" hypothesis alone is inadequate to account for hurricane formation". Even less adequate it must be to explain the further intensification of mature forms of TCs to the stage of super typhoons and hurricanes of the highest category. Thus, in (Lin et al. 2014), on the basis of the model proposed by the authors, estimates are given of the power of the vertical flow of enthalpy (the sum of apparent and latent heat) from the ocean to the atmosphere during the evolution of the Haiyan super typhoon (2013). They are given in comparison with the characteristic values of flows arising under the influence of other TCs. From (Lin et al. 2014, Figures 2,3) it follows that in the case of Haiyan TCs, the fluxes reached extreme values of about 1600 W/m^2 at the peak intensity (at characteristic values for the other TCs considered, of the order of 400–900 W/m^2). It is with these extreme values of ocean heat transfer that the authors (Lin et al. 2014) explain the extreme intensity characteristics of Haiyan TC (maximum wind speed up to 80 m/s). However, given a $V_{\max}$ value of 80 m/s and a radius of the eye wall of Haiyan TC of 50 km (Mori et al. 2014), it is easy to estimate the dissipative losses according to the K. Emanuel model (Emanuel 1999), which is 0.030 PW. At the above-mentioned enthalpy flux power of 1600 W/m^2, the total flux through the ocean area covered by a circle with a radius of 50 km (i.e. inside the wall of the TC) will be about 0.013 PW, i.e. half as much dissipative losses. If we take into account that the efficiency of a TC as a heat engine is a few percent (Palmén and Newton 1969), then the vertical energy flow from the ocean should be recognized as a negligible factor in the energy balance of a mature TC. This is, apparently, one of the most important reasons for very modest progress in the development of a prognostic criterion for the rapid intensification of a TC according to the "local" values of geophysical fields in the vicinity of its center (Kaplan et al. 2010).

An alternative source of TC energy is advection (horizontal transfer) of latent heat from the periphery to the center of the TC in the lower troposphere. The indirect role of the ocean as the main heat accumulator is also great here, since air humidity in the lower troposphere is highly correlated with SST (Petty 1990). However, this role is manifested on much larger scales than the size of a tropical cyclone, and, to a large extent, in the period of time preceding the formation of the mature phase of the TC.

It should be noted that latent heat advection, as the main factor in the evolution of mature TC forms, is already indicated by the authors of the earliest studies: "The essential dynamical features <of tropical cyclones> were first described by Ferrell (1856), who, following Espy, indicated that the violence and duration of the hurricane "depend upon the quantity of vapor supplied by the currents flowing in below"", (Palmén and Newton 1969, p. 472), see also (Ferrel 1856). If we assume that a powerful TC generates a stable convergent flow of air to the center of the vortex through a circular boundary with a radius $r = 8 \cdot 10^5$ m (the conditional radius of the outermost closed isobar of about 8°) with an effective (radial) speed $v = 2.5$ m/s at characteristic TPW values of $W = 50$ kg/m^2, then the corresponding latent heat flux power is given by

$$Q = q \cdot W \cdot 2\pi r \cdot v, \tag{4.1}$$

where $q = 2.26$ MJ/kg, and equals about 1.5 PW. This simplest estimate is in good agreement with the simulation results presented in (Palmén and Newton 1969, p. 498). So, for the "mean tropical cyclone" Palmén and Riel obtained the value of the total power of the latent heat flow through a circle with a radius of 2°, equal to 0.55 PV. At the same time, the power of the enthalpy flow from the underlying surface is estimated at 0.01 PW (55 times less), (ibid., Table 15.3). A somewhat smaller ratio of the contributions of the advective and convective components of the heat influx was obtained in other works when modeling the cases of two real TCs, however, for a circle with a radius of only 1°, (ibid., Table 15.4). In general, an unambiguous conclusion is made that "... the essential source of total <TC> energy is derived from the lateral influx of water vapor in the moist surface layer ...", (ibid., p. 504). It should be noted that modern work supports this conclusion with both new simulation results and analysis of satellite observation data (Trenberth et al. 2007; Sharkov 2012; Fritz and Wang 2014; Makarieva et al. 2017).

From the point of view of satellite remote sensing, the main problematic aspect of studying the advection factor in the analysis of the evolution of mature TC forms is the need to switch from "point", "instant" estimates to a detailed study of the changes in the two-dimensional spatial structure of the TPW field over time. This leads to the need to solve the inverse problem of restoring the dynamics of the geophysical fields of the atmosphere, discussed in the previous chapter, which is achieved in the developed approach of satellite radiothermovision.

4.2 General Characteristics of the Data Used and Analysis Methods

The book demonstrates two options for processing radiothermal satellite information. In the first approach (Ermakov et al. 2011, 2013a, b), the fields of radiobrightness temperatures on the initial SSM/I measurement grid were used as input information. The necessary data were taken from the Global Fields database (Ermakov et al. 2007). The calculation of the TPW field was carried out according to the method proposed in (Ruprecht 1996) (see Eq. (2.4) in Chap. 2). The disadvantage of this technique is the use of data from only two SSM/I measuring channels, and therefore, the accuracy of TPW recovery in conditions of dense cloudiness and precipitation is worsened (which, as a rule, corresponds to high values of TPW). It should be noted, however, that the TPW fields constructed using this method, when compared with those reconstructed using the method (Wentz 1997), exhibit an almost identical spatial structure, and, accordingly, can be effectively used to calculate atmospheric dynamics. On the other hand, artifacts of the methodology (Ruprecht 1996), manifested in the form of outliers against the background of high TPW values, form additional "tracers" that allow more efficient identification of the dynamics of areas that are characterized by uniformly high TPW values in the methodology (Wentz 1997), which prevents the restoration of their dynamics. At the same time, the errors in the calculation of integrated latent heat fluxes associated with these outliers do not seem to be very significant and are not able to qualitatively affect the analysis results. However, in order to more convincingly demonstrate this fact, the second processing option was also applied (Ermakov et al. 2015, 2016). In this version, the TPW fields already restored from satellite radiometry data were taken as initial data. The basis was the processing products of according to the procedure (Wentz 1997). However, their spatial resolution (0.25°) turned out to be unsatisfactory when analyzing the processes of rapid intensification of TCs, and therefore, these products were used as additions to the processing results of the AMSR-2 instrument (satellite GCOM-W1), which allowed spatiotemporal alignment and interpolation on the grid with a step of 0.125° according to the methodology (Ermakov et al. 2016) (see Chap. 3).

Thus, the calculation of the interpolated TPW fields was performed on regular grids of global coverage with a spatial step of 0.2 and 0.125°. The time interpolation step in these cases was 1.5 and 0.75 h, which ensured the calculation of the advection vector field with a time step of 3 h and 1.5 h, respectively (Ermakov et al. 2016). The use of synchronous and spatially combined fields of TPW and advection of water vapor, in turn, made it possible to calculate the total power of the advective flow of latent heat Q through families of concentric circular loops that surrounded the tropical cyclones under study and drifted with them during the evolution of the latter. Time series of the indicated Q values were constructed and compared with synchronous time series of intensity (maximum wind speed V) of the corresponding tropical cyclones for the entire period of their existence with the maximum possible time detail. Data on the trajectory and intensity of the TCs were obtained from (Pokrovskaya and Sharkov 2006, 2016).

Additional capabilities of satellite radiothermovision in the complex analysis of various satellite information are demonstrated by the examples of joint processing of the TPW fields and ocean surface temperature (SST) reconstructed from radiothermal and infrared satellite observations (Ermakov et al. 2015). In order to more correctly compare the fields of different geophysical parameters, their mutual spatiotemporal combination was carried out using the already described satellite radiothermovision algorithms. On the one hand, this made it possible to perform a synchronous analysis of the states of the ocean and the lower troposphere in dynamics during the passage and evolution of a tropical cyclone. On the other hand, the applied methodological framework obviously has significantly wider application areas, and can be used for spatiotemporal combination and complex analysis of a number of other fields of geophysical parameters (cloud liquid water content, precipitation rate, surface wind speed, etc.).

The methodology for calculating the total power of the latent heat flux through a circular contour is most adequate when analyzing the evolution of an individual tropical cyclone. However, the methodological basis developed and demonstrated in the book provides the possibility of similar studies of a wide range of mesoscale atmospheric processes. For this purpose, it is only necessary to setup a family of boundaries whose parameters (shape, size, displacement rate, possible deformations, etc.) are most consistent with the properties of the processes under study. In the book, the case of a system of interacting twin typhoons is considered, for which, while maintaining a common methodological approach, it was necessary to carry out calculations of fluxes through a family of more complex boundaries than concentric circles.

4.3 Evolution of Tropical Cyclones in the Field of Total Precipitable Water

This section briefly describes some examples of the practical application of the satellite radiothermovision approach for the analysis of convergent and divergent latent heat fluxes realized in the process of tropical cyclone evolution. A detailed discussion of the results is contained in (Ermakov et al. 2019a, b). The general idea in the considered examples was as follows. All tropical cyclones (TCs) operating in the time interval from July 28 to August 26, 2000 were investigated, see Table 4.1. Each TC was surrounded by a family of concentric circular contours of different radii, the time series of latent heat fluxes through these contours were calculated, which were then compared with synchronous series of the TC intensities (maximum sustainable wind speed in the wall). The calculations were performed for drifting contours; the drift velocity was estimated from (3.35) (see Chap. 3).

The robustness of the results of calculations of latent heat fluxes Q with respect to small changes in the integration contours was analyzed and, in general, confirmed (Ermakov et al. 2019a, b), provided that the TC is sufficiently far from the coastline. As the TC approaches large land masses, instability may arise due to less reliable estimates of the TPW values and the velocity of advection over some nodes of the computational grid. Instability manifests itself in the form of strong discrepancies in time series of Q values calculated for contours of different radii. In addition, one should pay attention to sharp fluctuations in the time course of Q that occur synchronously on time series corresponding to integration contours that are similar in size and sometimes lead to a change of Q sign to the opposite. These fluctuations are noted and discussed in (Ermakov et al. 2019a, b). The methodological complexity of

Table 4.1 Tropical cyclones of August 2000

N	TC name	Lifetime	Basin
1	Alberto	08/03–08/23	Northwest Atlantic ocean
2	Debby	08/16–08/25	ibidem
3	Bilis	08/15–08/24	Northwest Pacific ocean
4	Ewiniar	08/05–08/19	ibidem
5	Jelawat	07/31–08/12	ibidem
6	Giema	08/04–08/11	Northeast Pacific ocean
7	Hector	08/10–08/20	ibidem

the analysis lies in the fact that in the process of evolution of a TC its characteristic dimensions can significantly change along with the intensity. In this regard, the complex processes of redistribution of water vapor in the vicinity of the TC should be considered on a large set of contours, highlighting, as the main ones, contours of different sizes at different points in time. The development of an objective criterion for such a procedure for changing the dimensions of the integration contour is a specific problem. In the examples below, the objectivity of the analysis is achieved using fixed-size contours. Issues of using "deformable" contours are partially discussed using the example of an analysis of a system of interacting typhoons (Sect. 4.5). See also the Ewiniar TC example in this section.

The most interesting case in the considered time interval is the TC Alberto (August 3–23, 2000). The trajectory of the TC is shown in Fig. 4.1a. The colors encode different stages of the evolution of Alberto according to (Pokrovskaya and Sharkov 2006): TL and L (tropical low and low)—blue, TD (tropical depression)—green, TS (tropical storm)—orange, STS (strong tropical storm)—red, T (hurricane/typhoon)—dark red. During its evolution, the TC three times reached the stage

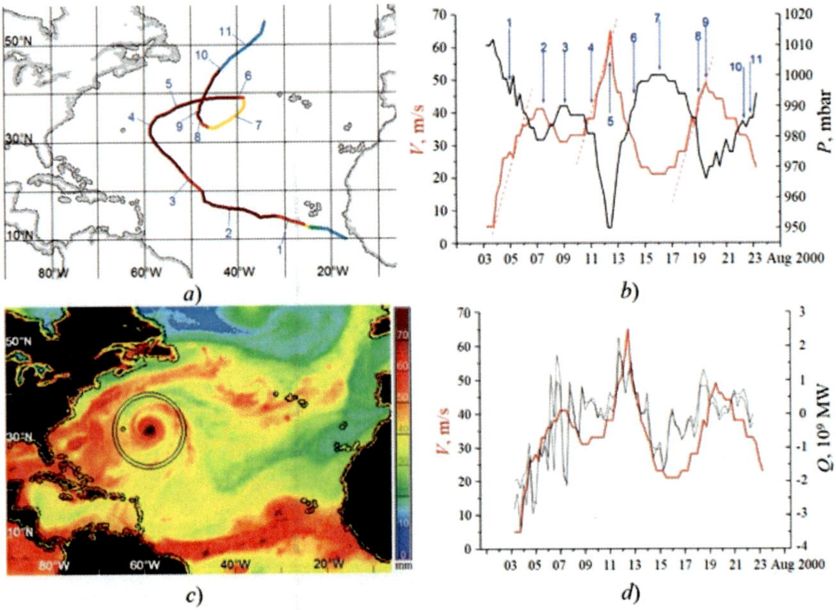

Fig. 4.1 Analysis of the evolution of TC Alberto: **a** the trajectory of the TC, color indicates different stages of development, geographical degrees are indicated along the edges of the figure; **b** the intensity of the TC (left scale) and the minimum pressure in the center of the eye (right scale), depending on the date in August 2000 (lower scale); **c** the TC in the TPW field and concentric circular contours of radii of 8 and 8.8° to calculate the power of the advective flux of latent heat; **d** the intensity of the TC (red line, left scale) and the power of latent heat fluxes (black lines, right scale) through predetermined contours, depending on the date in August 2000 (lower scale), see explanations in the text

of a hurricane. Figure 4.1b shows the evolution of the Alberto TC in terms of the maximum sustainable wind speed in the eye wall (red line) and the minimum pressure in the center of the eye (black line). As can be seen, the achievement of the hurricane stages each time was preceded by a phase of rapid intensification of the TC: a characteristic threshold value of ~15 m/s per day (Kaplan et al. 2010) is illustrated by the slope of the red dashed lines. The same numbers in Fig. 4.1a, b mark the same time points.

Figure 4.1c shows the TC in the TPW field at 11:00 UTC on 08/11/2000. The color scale in mm is shown on the right. The TC is surrounded by two contours with radii of 40 nodes of the computational grid (~8°) and 44 nodes (~8.8°). A pair of contours was used to verify the robustness of the flux calculations to small changes in the contour. As can be seen from Fig. 4.1a, c, throughout its existence, Alberto moved quite far from the coastline, which ensured better calculation quality: the proximity of land causes both an increase in the TPW errors and additional errors in the calculation of displacement vectors.

Figure 4.1d shows the calculated graphs of latent heat fluxes in Megawatts (MW) through the contours shown in Fig. 4.1c (black lines). To facilitate comparison, the intensity curve of the TC (maximum sustainable wind speed in the eye wall from Fig. 4.1b, red line) is reproduced in the same graph.

The curves of fluxes are everywhere close to each other, which indicates the robustness of the calculations with respect to small changes in the boundary contour. In addition, it can be noted that the curves well reproduce the main features of the TC evolution, namely, the intensification phase corresponds to a positive, convergent flow of latent heat (inside the contour), and the divergence to dissipation.

The first rapid intensification of the TC Alberto (an increase of about 15 m/s per day) led to its hurricane stage with a maximum wind speed V of more than 40 m/s on 08/07/2000 (Fig. 4.1b). The intensification was accompanied by a general increase in Q fluxes (with the previously noted fluctuations), which reached a maximum (near $1.0 \cdot 10^9$ MW) on the eve of 08/07/2000. This moment is marked in Fig. 4.1a, b by the number 2. Then, a slight weakening of the TC and Q fluxes followed, after which the second phase of rapid intensification began. The moment when Alberto reached the westernmost point of its trajectory is marked with a number 4 in Fig. 4.1. It can be seen that the integration contours (Fig. 4.1a) are located at a considerable distance from the coastline, which is also the case with all other Alberto positions. The second intensification ends with the achievement of the maximum strength of the TC on 08/12/2000 ($V \approx 65$ m/s), number 5 in Fig. 4.1. Just before this, the Q flux also reaches an absolute maximum near $1.8 \cdot 10^9$ MW (for the contours considered). Then weakening of the TC and a decrease in Q value (almost until the sign changes to negative) follow almost synchronously. Then, around August 16, 2000, a certain "surge" in values is observed, which coincides with the beginning of the third intensification of the TC, again leading to its reaching the hurricane stage with a maximum speed of about 49 m/s by the end of August 19, 2000 (Fig. 9 in Fig. 4.1). The maximum Q reached before this exceeded $1.0 \cdot 10^9$ MW. It is interesting to note that this last intensification took place at the intersection of the TC with its own "cold wake" (see the trajectory in Fig. 4.1a), which is clearly seen when studying the surface fields of the ocean surface, SST (see discussion in the next section). Thus, a rapid intensification of the TC was observed when passing over the ocean area, several days earlier (and at higher SST values) already involved in the interaction with the same TC without a significant change in the intensity of the latter.

The results of calculating the Q time course for the remaining TCs from Table 4.1 in comparison with their intensity are illustrated in a more concise form than in Fig. 4.1, in Figs. 4.2, 4.3, 4.4, 4.5, 4.6 and 4.7. Each of the figures shows the trajectory of the TC (dark gray line), as well as an example of integration contours (light circles) to visually assess the proximity of the TC to land masses. Separate graphs are given of the calculations of the Q time course (thin black lines, values in MW are plotted on the right scale) along with a graph of the TC intensity V (thick gray line, values in m/s are plotted on the left scale) for easy direct comparison. In the case of the TC Ewiniar, two pairs of contours are shown (light gray and black circles in Fig. 4.4a) and two corresponding calculation options (Fig. 4.4b, c).

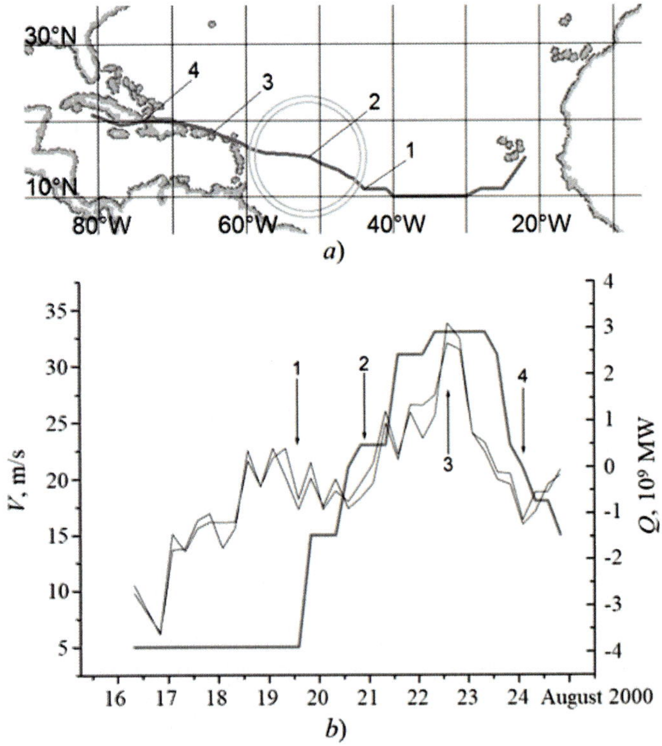

Fig. 4.2 TC Debby (North Atlantic): **a** TC trajectories and examples of integration contours; **b** the intensity of the TC (thick gray line, left scale) and the power of advective fluxes of latent heat (thin black lines, right scale), depending on the date in August 2000. See explanations in the text

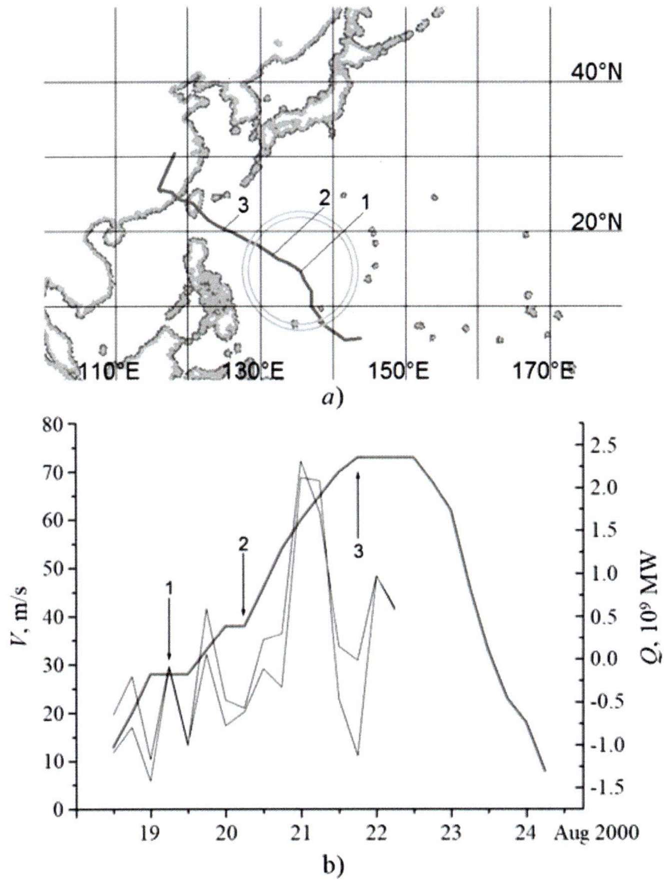

Fig. 4.3 TC Bilis (Northwest Pacific): **a** trajectory; **b** evolution

The numbers in the figures indicate some characteristic positions and stages of the evolution of a TC, the coincident numbers in each of the Figs. 4.2, 4.3, 4.4, 4.5, 4.6 and 4.7 correspond to the same time points.

Figure 4.2 shows the trajectory and evolution of the TC Debby (August 16–25, 2000). The TC moved along a straight line in the leading stream over the Central Atlantic and collapsed over Cuba (Fig. 4.2a). Its evolution is characterized by monotonous amplification to the hurricane stage (with a maximum wind of about 33 m/s), achieved on 08/22/2000, and the previous phase of rapid intensification (Fig. 4.2b). The plots of Q fluxes generally reflect the same evolutionary pattern. The initial stage of TC formation (TL according to (Pokrovskaya and Sharkov 2006, 2016)) is characterized by large divergent fluxes, which probably reflects the large-scale process of vapor transfer in the leading stream. The beginning of the actual development of the TC (from the stage of tropical depression, TD according to

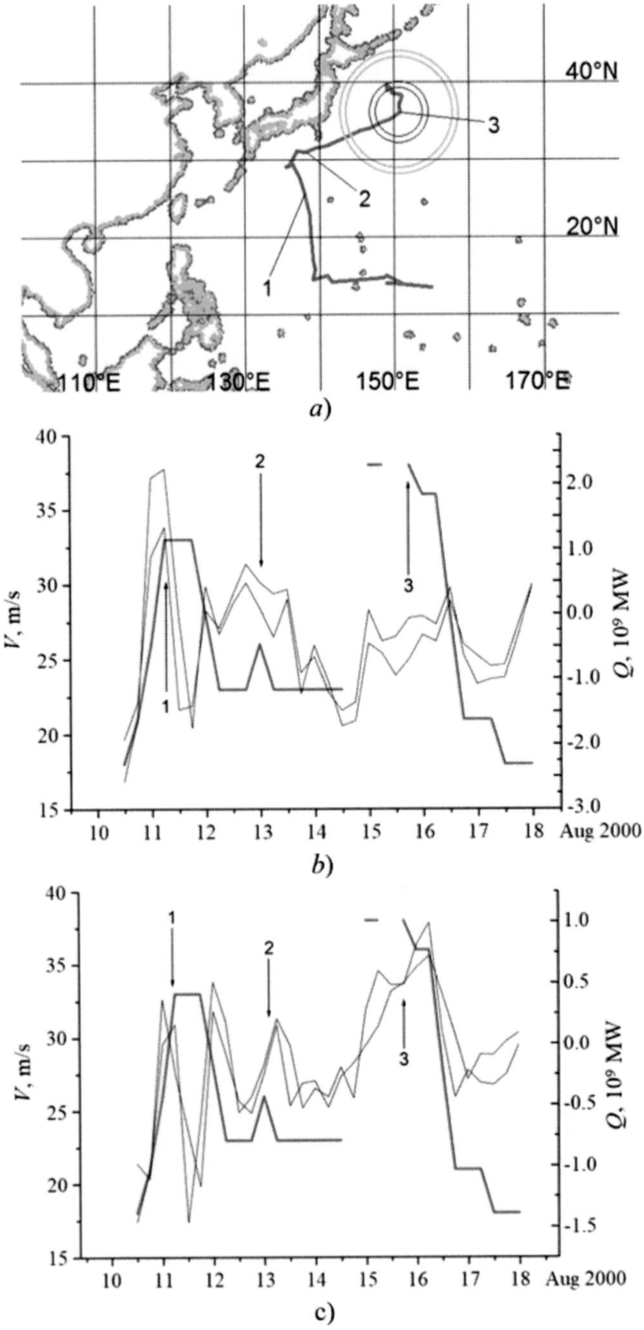

Fig. 4.4 TC Ewiniar (Northwest Pacific): **a** the trajectory of the TC and concentric contours with radii of about 4° (thin black circles) and 8° (thick gray circles); **b** the intensity of the TC and the power of the advective fluxes of latent heat through the contours of 8°; **c** the same with 4° contours

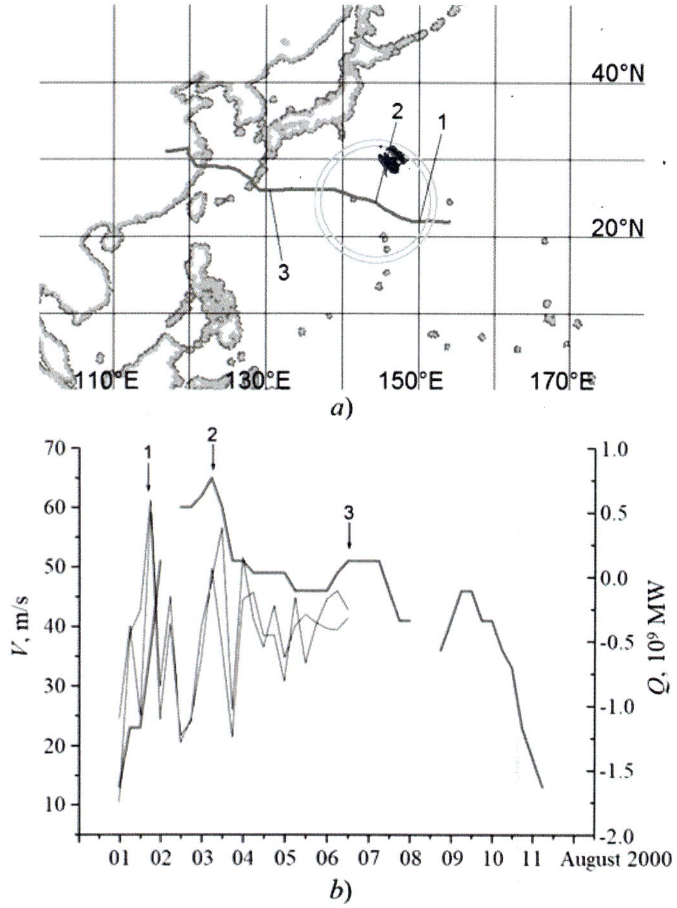

Fig. 4.5 TC Jelawat (Northwest Pacific): **a** trajectory; **b** evolution. See notes to Fig. 4.2 and explanations in the text

(Pokrovskaya and Sharkov 2006, 2016) with an increase of V up to 15 m/s) is indicated in Fig. 4.2 by the number 1. Directly before this, the first noticeable excess of the zero level fluxes (about $0.4 \cdot 10^9$ MW) was noted.

The middle of the rapid intensification phase is marked in Fig. 4.2 by number 2. In this phase of TC evolution, the fluxes increase to about $1.0 \cdot 10^9$ MW (Fig. 4.2b). It should be noted that at this point the TC approached the coastline, and the disturbing effect of land could begin to appear in the plots of the fluxes. This should be borne in mind when comparing the mutual time course of V and Q when analyzing the absolute estimates for Q: about $3.0 \cdot 10^9$ MW at a sharp peak, which occurred in the middle of the phase of the maximum power of the TC Debby (Fig. 4.2, number 3). As already noted, a deeper analysis of the evolution of a TC, especially when approaching the coast and with sharp changes in the intensity of the TC, is possible

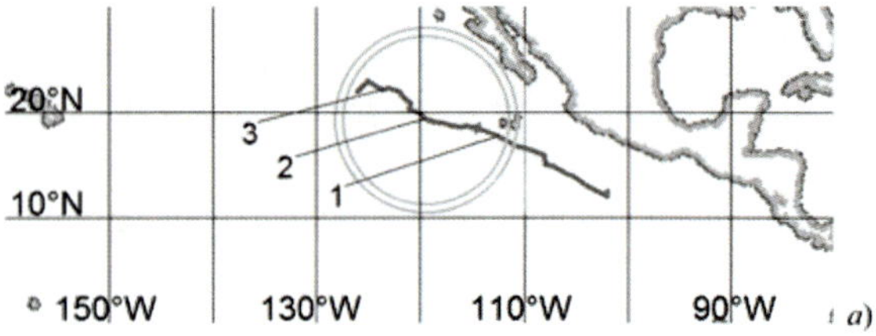

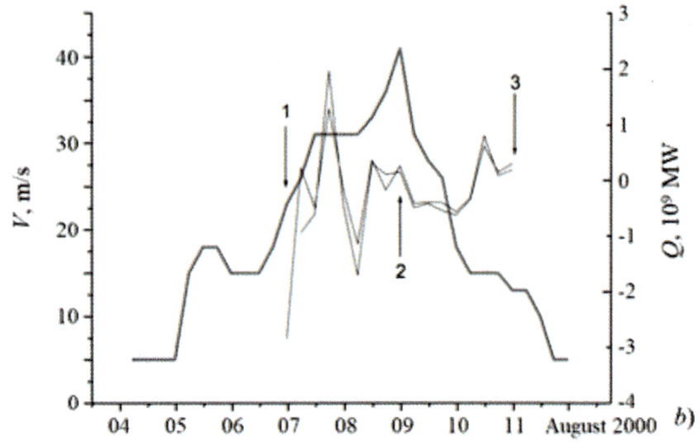

Fig. 4.6 TC Giema (Northeast Pacific): **a** trajectory; **b** evolution

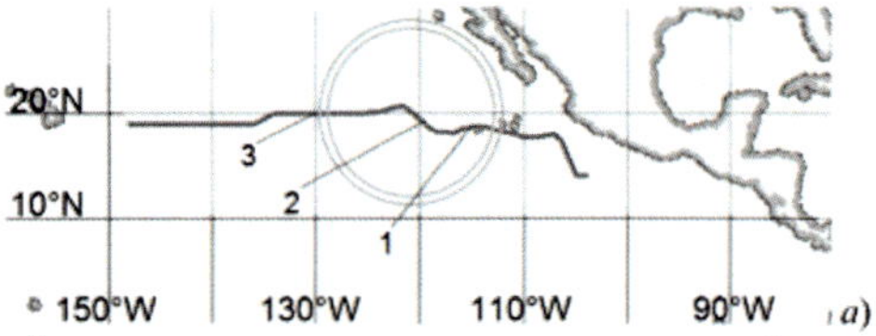

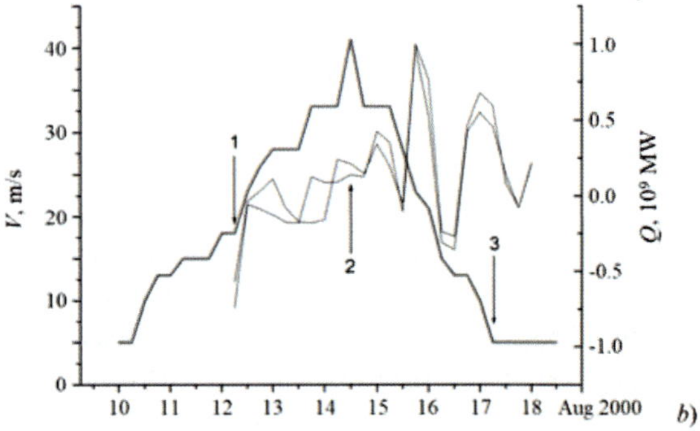

Fig. 4.7 TC Hector (Northeast Pacific): **a** trajectory; **b** evolution. See notes to Fig. 4.2 and explanations in the text

only with calculations on a large set of contours. In the phase of dissipation and destruction of the TC Debby that began after 08/23/2000, the fluxes sharply drop to 0, assuming substantially negative values at some points (number 4 in Fig. 4.2b).

Figures 4.3, 4.4 and 4.5 show the TCs Bilis, Ewiniar and Jelawat, which developed over the northwestern Pacific. From the point of view of the analysis performed, the Bilis and Jelawat TCs present cases similar to the TC Debby case considered above. Their trajectories do not contain complex features and lead in the west (north-west) direction from open water to land. The evolutionary chart as a whole is reproduced by the corresponding graphs, taking into account previously made remarks regarding fluctuations in the values and the influence of land proximity. So, at the beginning of the evolution of the TC Bilis, the fluxes fluctuate around 0 (number 1 in Fig. 4.3), reaching $0.5 \cdot 10^9$ MW before further rapid intensification. The phase of rapid intensification, the beginning of which is indicated by the number 2 in Fig. 4.3, falls on August 20–21, 2000, during which the fluxes exceed $2.0 \cdot 10^9$ MW at the maximum. The maximum strength (wind speed of about 73 m/s) of the TC Bilis is reached in the evening of 08/21/2000 (number 3 in Fig. 4.3). Soon the TC is approaching the Eurasian continent, and further calculation of fluxes becomes impossible.

Figure 4.5 shows a similar picture of the development of the TC Jelawat. The graphs of Fig. 4.5b started directly from the phase of rapid intensification of the TC, during which the Q fluxes reached an absolute maximum of $0.6 \cdot 10^9$ MW, and the wind speed V increased from 13 to 51 m/s in one day (number 1 in Fig. 4.5). The maximum strength of the TC was reached on 08/03/2000 (number 2 in Fig. 4.5), after which it began to slowly weaken, accompanied by a decrease in Q fluxes. As approaching to the coastline, the TC slightly increased several times, lasting until 11/08/2000, however, for the final part of its trajectory, starting from 08/06/2000 (number 3 in Fig. 4.5), calculations of fluxes are impossible due to the proximity of land.

Figure 4.4 illustrates an interesting case of the TC Ewiniar. This TC had a more complex trajectory, relatively remote from land, and dissipated over the ocean after two distinct phases of intensification (Fig. 4.4a). The fluxes analysis, as for the TC Jelawat, began with a phase of rapid intensification with growth of V from 18 to 33 m/s per day on August 10–11 2000 (number 1 in Fig. 4.4). It corresponds to a convergent Q flux of about $2.0 \cdot 10^9$ MW (Fig. 4.4b). Less than a day later, on 08/12/2000, the fast weakening of the TC begins, the Q fluxes fluctuate around 0, the wind speed drops to 23 m/s, remaining at this level for several days. The TC drifts northeast along the coast of Japan. A small jump of V (number 2 in Fig. 4.4b) is reflected in the plots of the Q fluxes, however, it should be noted that at this point the TC is most close to the coastline, which negatively affects the reliability of the estimates.

In the evening of 08/14/2000 or in the morning of the next day, the TC experiences a repeated rapid intensification. The data on the intensity of the TC for this day contain noticeable omissions (possibly due to the premature termination of its regular monitoring). The wind speed increases to 38 m/s and remains at this level during the day of August 15, 2000 (number 3 in Fig. 4.4), then the TC quickly weakens and collapses. The plots of the Q fluxes in Fig. 4.4 does not practically reflect this secondary intensification of the TC Ewiniar. As noted above, the study of the Q fluxes through one contour of a fixed size does not provide a sufficiently complete picture of the observed process of TC evolution. In the examples considered, such a limited-formal approach is adopted in order to minimize the subjective factor associated with an arbitrary choice of contour.

However, in the case of the TC Ewiniar, there are objective prerequisites for considering contours of a smaller radius. First, the indicated TC was approximately half as intense as the TC Alberto, which was initially considered (see Fig. 4.1). Secondly, it is in the case of the TC Ewiniar that the choice of the radius of the contours of 8° is unlucky, because when this contour is displaced along the second half of the TC trajectory (the interval between the numbers 2 and 3 in Fig. 4.4a), its edge slides over the coast of Japan, which reduces the reliability of the estimates. For these reasons, for the TC Ewiniar, contours of half the smaller radius are additionally considered. In Fig. 4.4a, the "standard" contours are shown by light gray circles (the graphs in Fig. 4.4b correspond to them), and the reduced contours are shown in black circles (the graphs in Fig. 4.4c correspond to them).

Comparing Figs. 4.4b, c, it should be noted, first of all, that they similarly reproduce the nature of the flux changes for the first half of the trajectory (up to the number 2 in Fig. 4.4), differing only in amplitude (for internal contours, the maximum Q values are much smaller). At the same time, however, the graphs of Fig. 4.4c satisfactorily reproduce the evolution of the TC Ewiniar in the final part of its trajectory, reaching the values of $1.0 \cdot 10^9$ MW in its maximum intensity (number 3 in Fig. 4.4). This example once again emphasizes the importance of calculating and analyzing fluxes from a complex set of contours when studying the evolution of a TC.

Figures 4.6 and 4.7 illustrate cases of the TCs Giema and Hector of the northeast Pacific. For them, as well as for other TCs in this water area, formation and movement near the coast of Central and South North America, intensification in the process of removal into the open ocean with dissipation far from land, are characteristic. Although the physical mechanism of development of these TCs is undoubtedly the same as in all other cases, their study using the methodology adopted here is greatly complicated by the initial proximity of the TCs to the mainland. On the other hand, the relatively low risk of damage from these TCs for the coast and inland areas of North America makes TCs less important in terms of practical applications.

For the two TCs considered (Giema and Hector), it is characteristic that, at the moment of reaching maximum strength, each of them was not yet far from the American continent so that the flux estimates be sufficiently reliable (see Figs. 4.6a, 4.7a). Therefore, the graphs in Figs. 4.6b and 4.7b are given without detailed analysis to complete the review of the TCs developed in August 2000. Both TCs, especially the TC Hector, should be considered unsatisfactory objects for studying the relationship between the evolution of the TC and the time course of the Q fluxes, because the latter cannot be reliably calculated due to significant errors in estimates of both the TPW field and the advection rates near the continent.

So, based on the approach of satellite radiothermovision in analyzing the evolution processes of TCs in 2000, the relationship between changes in the intensity of existing TCs and variations in latent heat fluxes through the contours surrounding them is revealed in the TPW field. This relationship is most clearly seen for the TCs moving at a considerable distance from the continents (which is explained by methodological limitations, and not by the nature of the phenomenon). Convergent latent heat flux (to the eye of the TC) corresponds to the phase of its intensification and reaches values of the order of $1.0 \cdot 10^9$ MW or more in total power. The divergent flux corresponds to the dissipation phase of the TC. In particular, the graphs of the time course of the calculated latent heat fluxes well reproduce the picture of triple rapid intensification of the TC Alberto, twofold rapid intensification of the TC Ewiniar, as well as the evolution of the TCs Bilis, Debby and Jelawat.

It should be emphasized the importance of in-depth analysis based on estimates of latent heat fluxes through a large set of different contours, including taking into account the current intensity of the TC and its proximity to large land masses. Such an analysis is also important for the study of cause-effect relationships through a sequence of phases of the observed processes. On the one hand, the inflow of latent heat to the area of the eye of the TC leads to its intensification, on the other hand, the intensification of the TC under favorable conditions (high TPW of the surrounding atmosphere) is able to organize and enhance convergent fluxes of latent heat in an even larger region of the troposphere.

4.4 Complex Analysis in the Fields of Several Geophysical Parameters

In this section, the calculations of the time series of the power of the advective fluxes of latent heat are supplemented by the calculations of the series of ocean surface temperatures (SST) along the trajectory of the studied tropical cyclones (TC) immediately before and after their passage. In addition to the previously considered TC Alberto, an important example of an analysis of the evolution of the super typhoon Haiyan (2013) and the subsequent tropical storm Podul (which did not reach the hurricane phase) is presented here. The methodology for calculating latent heat fluxes repeats that described in the previous sections. The only difference from the previously considered cases was that the processing was not started from the level of the field of radiobrightness temperatures. Instead, RSS processing products (Wentz et al. 2012) (for data from SSMIS F16, F17) and JAXA (for data from AMSR-2 GCOM-W1) were used, from which the total precipitable water (TPW) fields were extracted. In the case of RSS products, TPW fields were initially constructed on a global regular grid with a step of 0.25°. The TPW fields from JAXA processing products were interpolated onto the grid with a step of 0.125° according to a procedure similar to that described in Chap. 3 (construction of reference fields). Further, using the multisensory approach (Sect. 3.5 of Chap. 3), the TPW fields for all types of measurements were combined, as a result of which interpolated fields of total global coverage were obtained on a regular grid with a step of 0.125° and a time step of 1.5 h. At the same time, during the spatiotemporal interpolation, vector advection

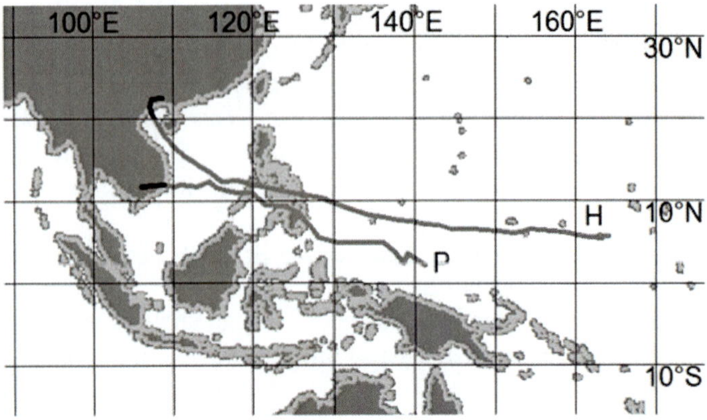

Fig. 4.8 Trajectories of the TC Haiyan (H) and TS Podul (P). Geographic coordinates are indicated in degrees north (N) and south (S) latitude and east (E) longitude

fields were also used, as before, to calculate latent heat fluxes through the families of concentric contours surrounding the TCs.

For analysis of SST, daily composite product MW OI SST (Microwave OI SST … 2020; Gentemann et al. 2003) by Remote Sensing Systems were used. For each position of the eye of the TC along the trajectory of its movement, the average SST values were calculated inside a circle with a diameter of 1.5° one day before passing of the TC above this place, as well as one day and three days after the TC passage. This made it possible to analyze all the anomalies that occurred in the SST field during the evolution of the TC and its movement over the ocean.

Super typhoon Haiyan was one of the most interesting cases of the TCs in November 2013 (Mori et al. 2014). It formed over the northwestern part of the Pacific Ocean over deeply warmed ocean waters (Lin et al. 2014) and, on 11/02/2013–11/11/2013, it quickly moved to land almost strictly westward along a rectilinear trajectory, Fig. 4.8. The fast drift of the Haiyan TC is considered one of the factors that determined its extreme intensity parameters due to the relatively small friction losses on the roughened surface of the ocean (ibid.). This, however, suggests that the upper active layer of the ocean was practically not involved in the thermodynamic interaction with the TC. Indeed, in the SST field there is no pronounced "cold wake" behind Haiyan, and in the time series of the SST constructed by the method described above, there are no significant anomalies, Fig. 4.9a. It can be concluded that the influence of the TC Haiyan on the upper active layer of the ocean (in terms of heat transfer in the ocean-atmosphere system) turned out to be relatively small. Nevertheless, several tropical storms that followed it, including Podul, formed just a week after Haiyan (11/09/2013–11/15/2013) and moving along a parallel path along equally warmed ocean waters, Fig. 4.8, did not reach hurricane phase.

Figure 4.9a illustrates the time course of the TC Haiyan intensity, V, m/s (gray line, right scale of values) and the SST value at the points of its trajectory one day before (black squares), one day (white circles) and three days after (white triangles) of its

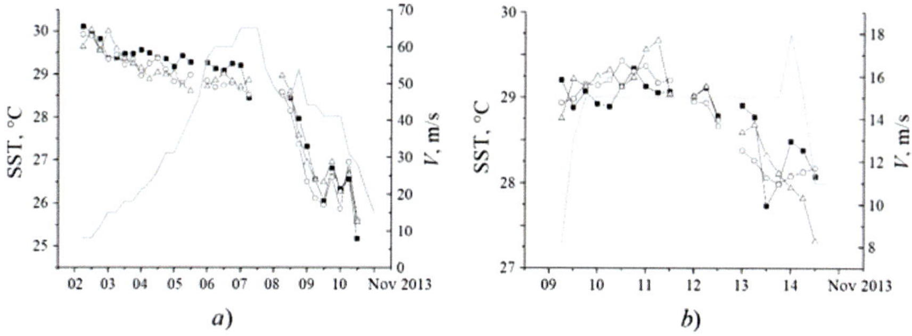

Fig. 4.9 **a** the intensity of the TC Haiyan (right scale) depending on the date in November 2013 at local noon (bottom scale) and the SST value (left scale) at the corresponding points of the TC trajectory (squares—one day before the TC passing; circles—one day after the TC passing; triangles—three days after the TC passing); **b** the same for the TS Podul

passage. It can be seen that the change in SST does not exceed 0.5–1 °C modulo anywhere (moreover, the difference can be either negative or positive in different parts of the trajectory). In the intensification phase of Haiyan, SST was 29–30 °C. Very close SST values (in any case, significantly higher than the temperature of 26 °C, the conditional phenomenological threshold for the generation of TCs) are also characteristic of the case of the tropical storm Podul, Fig. 4.9b. The absolute value of the SST differences before and after the passage of the storm also lies in the range of 0.5–1 °C, and the sign of the difference is different in different parts of the Podul trajectory. It can be seen, however, that moving for about two days (from 11/10/2013 to 12/12/2013) over warmed ocean waters until reaching the island of Mindanao (Figs. 4.8, 4.10b), Podul maintained an intensity of about 15 m/s, without passing into the hurricane stage. In a nearby section of the trajectory, the TC Haiyan

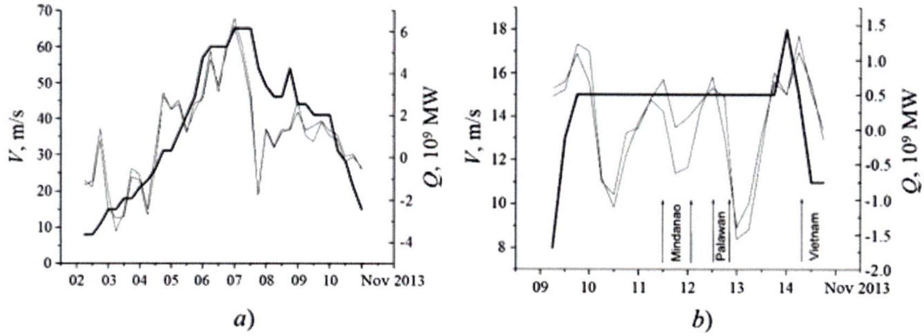

Fig. 4.10 **a** Evolution of Haiyan: a thick black line is the intensity of the TC (left scale), thin gray lines are the powers of advective fluxes of latent heat (right scale) through concentric circular contours with radii of 8 and 8.5° depending on the date in November 2013 at local noon (bottom scale); **b** the same for the TS Podul (radii of circular boundaries 4° and 4.5°, respectively); additionally indicated are the time intervals for the passage of the storm over land (Mindanao and Palawan Islands; Vietnam)

increased its intensity from 30 to (at least) 65 m/s in two days, i.e. passed the stage of rapid intensification (Figs. 4.8, 4.10a).

This nature of the evolution of two atmospheric catastrophes is also clearly visible in the graphs of advective latent heat fluxes Q calculated according to the method described above through the circular contours following the TC Haiyan (Fig. 4.10a) and the tropical storm (TS) Podul (Fig. 4.10b).

Figure 4.10a illustrates the time series of the TC Haiyan TC, V, m/s (thick line, left scale) along with the Q time series calculated for contours with radii of about 8° (see the previous section). The convergent nature of the fluxes is clearly visible, corresponding to the phase of intensification of the TC with a maximum of about 6 PW (right scale of values) and a further decline and transition to the divergent mode of flows. A similar picture is observed for contours of smaller radii.

In analyzing the evolution of the Podul tropical storm, smaller contours (with a radius of about 4°) were used both because of the proximity of its trajectory to land and because of its significantly lower intensity (see also the example of Ewiniar TC in the previous section). The picture of advective fluxes here is fundamentally different: all the time the storm has existed, they oscillate around zero with a relatively small amplitude of the order of 1 PW and experience a small surge with a slight intensification of Podul immediately before entering the coast of Vietnam. This correlates with the general smooth course of the Podul intensity (some of the nuances of which could not be reproduced in the data of the monitoring centers due to the low frequency of observations—not more than 4 times a day—and low accuracy, up to several m/s).

To conclude this section, let us once again consider one of the characteristic stages of the evolution of the TC Alberto, additionally using synchronized data on SST based on composite RSS products (see above). As can be seen from Fig. 4.1, the TC Alberto crossed its own trajectory on August 19–20 2000. The SST field immediately before this point (on 08/18/2000) is shown in Fig. 4.11a. The "cold

Fig. 4.11 **a** SST field in the North Atlantic (the color scale of temperature intervals with a step of 1 °C at the bottom of the figure), the trajectory of the TC Alberto (red line) and the position of the TC (red circle) on Aug 18, 2000; along the edges of the image—geographical coordinates; **b** plots of the TC intensity V, m/s (red line, left scale), average SST under the TC eye (green line, right inner scale) and the power of the advective flux of latent heat through the 8° boundary (black line, right outer scale); on the lower scale—dates in August 2000

wake" formed by Alberto during its passage over this region of the ocean between 08/13/2000 and 08/15/2000 (numbers 5–6 in Fig. 4.1a) is highlighted. Although its formation can be interpreted as a sign of intense heat exchange between the ocean and the TC, it should be noted that the TC Alberto on this day significantly weakened (see Fig. 4.1b). At the moment when the TC Alberto crossed its own trajectory, the SST was only about 23 °C. Nevertheless, Alberto was in the phase of the third rapid intensification (Fig. 4.1b, numbers 8–9, Fig. 4.11b). Thus, it is difficult to connect this phase of its evolution with the local state of the ocean. On the other hand, the time course of the total power of convergent fluxes of latent heat, as noted above, completely reproduces the picture of the time course of its intensity, Figs. 4.1d, 4.11b.

Thus, the evolution of the mature phases of the TC is determined mainly by the advection of latent heat from the surrounding atmosphere. Convergence of latent heat increases with increasing intensity of the TC and leads to its further intensification. Under less favorable conditions (with a lower TPW of the atmosphere, when the vortex is weakened, in particular, when it goes to land), the convergent flux of latent heat decreases and can be replaced by divergence, which leads to further weakening of the vortex.

4.5 Expanding the Approach for Exploring a System of Interacting Typhoons

The season of 2015 was characterized by high activity of tropical cyclogenesis, especially in the northern hemisphere, where 66 TCs were formed with a climatic norm of 57.2 (according to the Russian Hydrometeorological Center, https://meteoinfo.ru/categ-articles/33-climate-cat/monitoring-klimata/tropcyclones/12045–220155-). Under these conditions, the formation of systems of double (interacting) TCs is most likely, the frequency of occurrence of which is maximum in the northwestern Pacific region (Dong and Neumann 1983). A pair of typhoons Goni and Atsani (August 11–26, 2015), which for a long time moved in parallel courses over the open ocean, almost simultaneously changing their intensity, was considered as a candidate for the role of a system of twin TCs. Goni reached the Philippines on August 20 and, causing significant damage and casualties, the next day abruptly turned to the northeast. Almost simultaneously with it, Atsani made a similar turn over the open ocean and, having passed the peak of intensity, went into the dissipation phase, and finally collapsed following Goni after August 26, Fig. 4.12.

The systems of twin TCs attract increased attention of researchers due to the fact that the interaction of TCs can lead to significant deviations of their trajectories from the predicted ones and affect the evolution of the TCs. The most striking manifestation of this interaction is named after the author of the first experimental works in this direction (Fujiwhara 1921, 1923) by the Fujiwara effect and consists in the mutual attraction of two close vortices, leading to their rotation around a common center with possible subsequent merging. Further modeling showed (Dritschel and Waugh

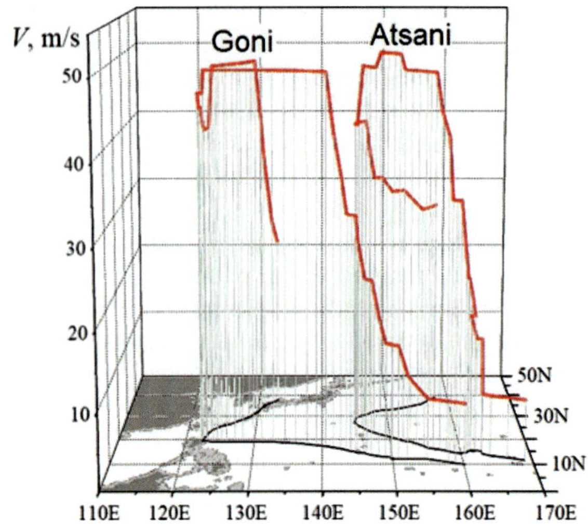

Fig. 4.12 Evolution of the TCs Goni and Atsani: trajectories (black lines) and intensities (red lines, vertical scale). See explanations in the text

1992; Falkovich et al. 1995; Prieto et al. 2003), that when two TCs interact, depending on the initial conditions, fundamentally different scenarios of joint motion can be realized: cyclical rotation with runaway, deformation or complete destruction of one TC, partial or complete merger of the TCs.

In (Brand 1970), based on statistical analysis, an estimate is given of the maximum distance between TCs (750 nautical miles or about 1400 km), at which the interaction of the TCs will noticeably affect their trajectories. A pair of Goni and Atsani formally satisfy the requirements of (Brand 1970) for the selection of systems for interacting TCs:

(1) at some point in time, their centers were at a distance of less than 700 nautical miles (about 1300 km), Fig. 4.13;
(2) TCs developed over the open ocean, Fig. 4.12;
(3) at some point both the TCs reached the typhoon stage, Figs. 4.12, 4.13.

However, it should be noted that for almost the entire time Goni and Atsani were increasingly distant from each other and did not show mutual cyclonic rotation (except for the time interval of August 18–22 at a distance of about 2000–2400 km from each other). Of course, the maximum interaction distance depends on the development conditions, the size and intensity of the vortices. So, in (Ziv and Alpert 1995) it is stated that extratropical cyclones, in some contradictions with two-dimensional models, can be attracted at a distance of up to 2000 km, but the estimate of 1400 km remains valid for TCs. Therefore, the trajectory features of Goni and Atsani should probably be explained primarily by the nature of the general circulation (Hoover 1961) and the state of large-scale geophysical fields in the region.

On the other hand, it is obvious that the interaction of the TCs cannot be limited only by the mutual influence on the trajectory of movement. An important energy aspect of the interaction of TCs is the formation of advective fluxes in

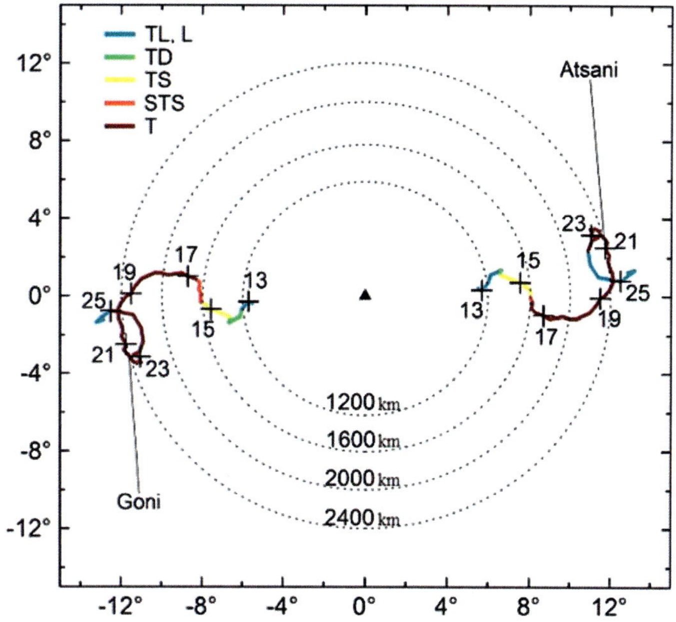

Fig. 4.13 The movement of the TCs Goni and Atsani relative to the geometric center of the system. The "+" symbols mark the beginning of the day (in UTC) on August 13–25, 2015, respectively. The phases of the TCs (TL, TD, TS, STS, T, L) are given according to (Pokrovskaya and Sharkov 2016)

the lower troposphere, as the main factor in the energy balance of TCs (Palmén and Newton 1969). Animation of cloud systems observed from Himawari-8, JMA meteorological satellites (www.jma.go.jp/en/gms/index.html?area=6&element=1&mode=UTC), SNPP, NOAA (www.nsof.class.noaa.gov/saa/products/search?datatype_family=VIIRS_IPGD), as well as visualization of the results of modeling the lower atmosphere (http://cimss.ssec.wisc.edu/goes/blog/archives/19282) show that the formation of air flows had a complicated character with a number of specific features, see Fig. 4.14. A zone of strong divergent fluxes to the south of the TC Goni is identified, indicated by the letter A in Fig. 4.14. Air flow into the system of the TCs was carried out from the south-west, then the flow branched, partially included in the cyclonic rotation around Goni (branch 1) and Atsani (branch 2). The area between the centers of the TCs, indicated by the letter B in Fig. 4.14, with low wind speeds and an unstable flow direction is also highlighted. As a result, a powerful flow branch was formed (branch 3), which surrounded the entire TCs system in a cyclonic direction.

This provides a basis for studying the evolution and energy balance of the system of TCs Goni and Atsani as a whole, depending on the magnitude of the advective fluxes of latent heat coming from the surrounding atmosphere. A general approach to the analysis of TC evolution in the TPW field is described earlier in Sect. 4.3.

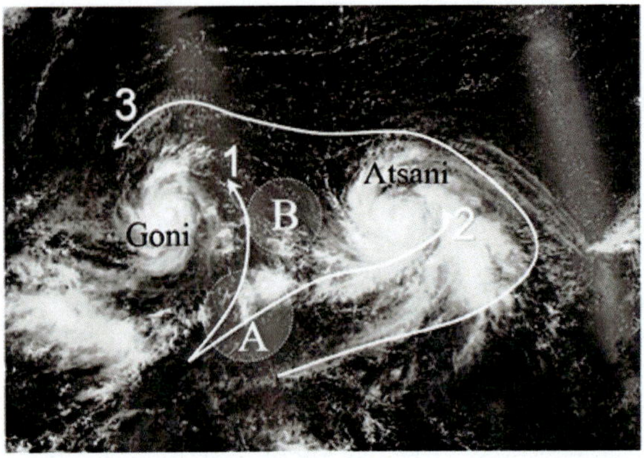

Fig. 4.14 Scheme of the main tropospheric flows in the system of twin TCs Goni and Atsani: A—region of strong divergent flows branching to the centers of Goni (1) and Atsani (2); B is the region of weak unstable flows; 3-stream covering the TC system in the cyclonic direction. The background image is based on NOAA archival data (www.nnvl.noaa.gov/view/globaldata.html#TRUE) on August 18, 2015

The primary source of data on TCs (position, intensity) was the Global-TC database (Pokrovskaya and Sharkov 2016). The global TPW fields on the 0.125° grid were built on the basis of standard satellite data processing products AMSR-2 (http://suzaku.eorc.jaxa.jp/GCOM_W/data/data_w_dpss.html), SSMIS and WindSat (http://www.remss.com/measurements/atmospheric-water-vapor) and interpolated in increments of 1.5 h. In this case, the multisensor algorithm for satellite radiothermovision was used (Ermakov et al. 2016). Then, based on the interpolated TPW fields and reconstructed advection fields, latent heat fluxes through the contours surrounding the system of twin TCs were calculated for comparison with the time course of the TC intensity.

The cases of analysis of the evolution of individual TCs are considered above. In these cases, contours in the form of a family of concentric circles moving together with the center of the TC were used. In this case, for the reasons indicated in the description of Fig. 4.14, such a choice of boundaries is not quite adequate. It should be expected that the complex nature of air currents, especially in the areas indicated by letters A and B in Fig. 4.14, will cause jumps of different signs in the fluxes formally calculated through circular contours, which will reflect not so much the energy gain or loss by the TC system as its spatial redistribution inside the system (isobar deformation). This assumption was verified by calculating the fluxes separately for the TCs Goni and Atsani through the circular contours of relatively small radii.

Thus, the specifics of the problem were determined by the need to analyze the evolution of the system of TCs Goni and Atsani as a whole, which is associated with overcoming two methodological difficulties. The first was to numerically characterize the intensity of the TCs system. As it is known, the intensity of the TC (maximum wind speed V, m/s) is not an additive measure of its power, P, MW. Various types of relation between P and V are considered in the literature, from quadratic, as, for example, in the Accumulated Cyclone Energy Index, ACE (http://www.cpc.noaa.gov/products/outlooks/background_information.shtml) to more complex (Emanuel 2005), which is partly due to terminological differences in determining the power of the TC. In the case considered, the situation was significantly simplified due to a similar quasi-synchronous picture of the evolution of Goni and Atsani and similar values of their intensities V_G and V_A at all times. Values proportional to various measures of the total intensity of the TC, for example, $V_1 = \frac{1}{2}(V_G + V_A)$ and $V_2 = \frac{1}{\sqrt{2}}\sqrt{V_G^2 + V_A^2}$ practically coincide with the corresponding normalization, Fig. 4.15. Therefore, for a qualitative description of the evolution of the system in order to compare with the advective fluxes of latent heat, any of the indicated values can be used (hereinafter V_2 is used).

The second difficulty was the choice of contours enclosing the TCs system. It is clear that in this case, the contours must have a complex, different from a circular shape. The calculation method in this case does not present a significant problem, however, the question arises of the justification of the choice of one or another contour, which during the evolution of the system must undergo various deformations, not excluding significant changes in size. An analysis of approaches to such a justification is the subject of further research. It may be necessary to attract significant amounts of additional information, for example, on pressure fields. Then the problem arises of obtaining or interpolating such data with spatiotemporal discretization satisfactory for joint analysis on a 0.125° grid with a time step of no more than 3 (better than 1.5) hours (Ermakov et al. 2017).

In the case considered, the following simple approach to the formation of the contour turned out to be quite effective. Two circular contours were constructed, centered relative to the Goni and Atsani eyes, respectively, with radii large enough to ensure their mutual intersection on most of the TCs trajectories, Fig. 4.16. Next, the powers of the advective fluxes of latent heat through each of these contours Q_1 and Q_2 were calculated, and the results were added together. Assuming that the fluxes through the fragments of the contours between the points of their intersection

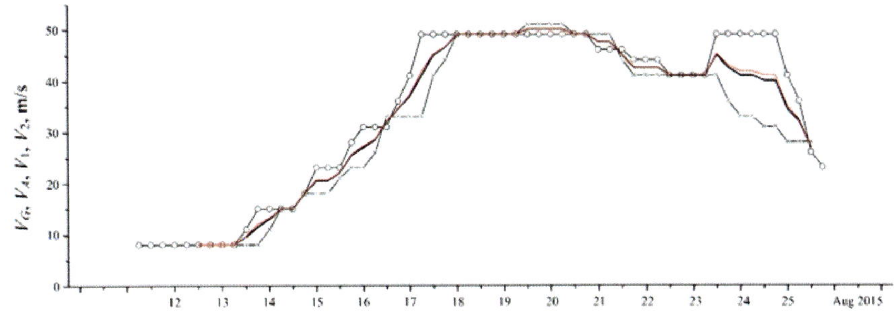

Fig. 4.15 The time course of the intensity (m/s) of Goni, V_G (circles, a thin black line), Atsani, V_A (diagonal crosses, a thin black line), as well as the normalized characteristics of the total intensity of the system of the two TCs: linear, V_1 (thick black line), and quadratic, V_2 (thick red line)

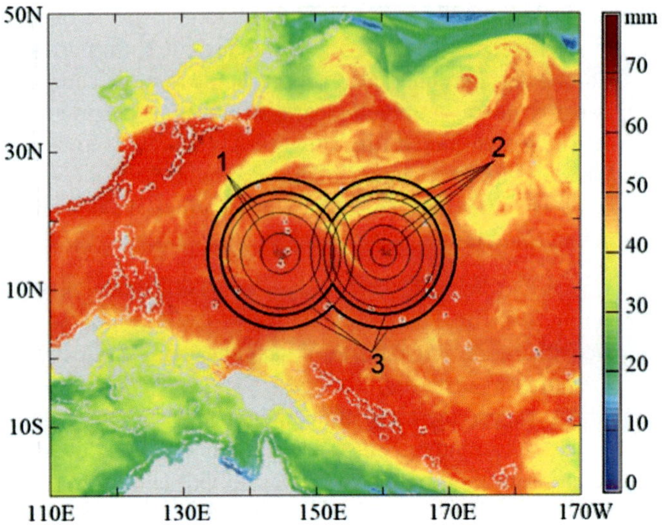

Fig. 4.16 Goni and Atsani in the TPW field (color scale on the right) and the families of contours considered: 1—around the TC Goni with r = 3°, 6°, 8°; 2—around the TC Atsani with r = 2°, 4°, 6°, 8°; 3—composite with R = 9°, 11°

during addition cancel each other out, we can assume that the sum $Q = Q_1 + Q_2$ characterizes the flux through the "external" parts of the contours. Thus, in fact, it was proposed to implement a composite contour that deforms and moves with the TC system and runs all the time the system exists at approximately the same distance from the nearest TC. The sums of fluxes Q were calculated for several R values and a further comparison of the time course of Q and intensity characteristics V_2 of the TCs system was performed.

The time course of the intensity of the TCs Goni, V_G, Atsani, V_A, and normalized characteristics V_1, V_2 of their total intensity, is shown in Fig. 4.15. The following main stages of evolution of the considered TCs system can be noted:

(1) intensification during August 13–18 (rapid intensification on August 17);
(2) the achievement of the first maximum intensity on August 20;
(3) some weakening with a minimum of August 23;
(4) re-intensification with a maximum of August 24;
(5) dissipation and final destruction by August 26–27.

The time course of the intensity of the TC Goni in comparison with the advective fluxes of latent heat through the contours surrounding its eye with radii of 3, 6, and 8° is illustrated in Fig. 4.17a. In contrast to the previously considered cases of analysis of individual TCs, the relationship between the intensity of Goni, V_G and heat advection through a contour with a radius of 8° (curve 4 in Fig. 4.17a) is relatively weakly expressed. It is possible to note the maximum of convergent flows around August 20, which coincides with the phase of maximum V_G values, and the divergent nature of the fluxes when the TC is weakened on August 21–23.

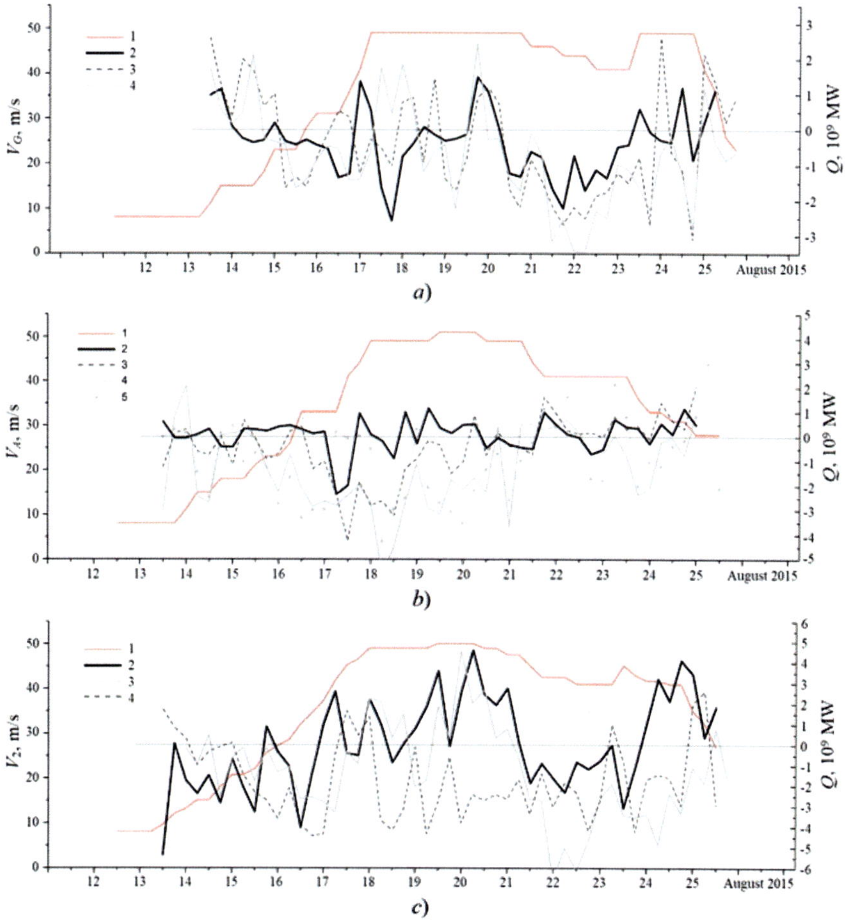

Fig. 4.17 Evolution of the Goni and Atsani system in comparison with latent heat advection: **a** Goni intensity, V_G (line 1, left scale) and latent heat advection, Q (lines 2–4, right scale) through the contours around the TC with radii of 3°, 6°, 8°, respectively; **b** Atsani intensity, V_A (line 1, the left scale) and advection of latent heat, Q (lines 2–5, the right scale) through the contours around the TC with radii of 2°, 4°, 6°, 8°, respectively; **c** the intensity of the Goni and Atsani system, V_2 (line 1, left scale) and the latent heat advection, Q (lines 2–4, right scale) through the composite contours around the system (see the example of circuits in Fig. 4.16), at $R = 11°, 9°, 8°$

However, negative Q values attracted attention during the rapid intensification of the TC on August 17–18 and an estimate close to 0 on August 24. Perhaps the best correspondence to the time course of V_G is shown by the flux calculated through a contour with a radius of 3°, however, comparison of estimates through all contours does not give a robust picture.

An even more controversial situation is revealed by an individual analysis of the TC Atsani, Fig. 4.17b. In this case, fluxes through circular contours with a radius of 2, 4, 6, and 8° are shown. The picture of convergent and divergent flows is highly randomized. Noticeable convergent fluxes are detected only with a contour with a radius of 2°; as the size of the contours increases, the estimates are shifted to the negative domain.

The picture fundamentally changes when analyzing the system of Goni and Atsani with the help of composite contours, Fig. 4.17c. Thus, a comparison of the intensity characteristics V_2 of the TCs system (curve 1) with the advection of latent heat Q through the composite contour of $R = 11°$ (curve 2, about 5 PW at maximum) reveals a clear correspondence in all phases of the system's existence. A close picture (with the exception of the last dates of August, 24–25) was also obtained for contour of $R = 9°$ (curve 3). However, already with $R = 8°$ this correspondence, especially in the first half of the evolutionary path of the system is sharply broken, which is probably due to the early opening of the contours and the increasing influence of the areas A and B problematic for analysis (Fig. 4.14).

It should be added that in the final stages of the existence of Goni and Atsani, they were separated by a distance of about 2400 km, and their analysis, as a system of twin TCs, with the chosen simplified approach ceases to be effective, including due to the disturbing effect of land, which generally increases with large radii. It can be noted that the final phases of the evolution of the TCs are satisfactorily described in their individual analysis by estimates through circular contours with a radius of about 6° since August 21.

In the case of the TC Goni (curve 3 in Fig. 4.17a), it is the decrease of Q from 1 PW to −2 PW in the phase of dissipation, an increase to 3 PW in the phase of re-intensification and a new decrease with fluctuations around 0 in the phase of final dissipation. In the case of Atsani (curve 4 in Fig. 4.17b), it is an alternation of convergent and divergent modes with a common negative trend of Q and a maximum of about 0.7 PW in the dissipation phase.

Thus, the satellite radiothermovision approach was first applied to analyze the evolution of a system of twin TCs. In this case, the interaction of the TCs in the system did not explicitly affect the trajectory features, but formed complex advective fluxes of latent heat in the lower troposphere. The calculation of these fluxes on the basis of a previously developed technique had an important feature related to the choice of enveloping contours of complex configuration. Due to the specific conditions of the development of the system of Goni and Atsani the approach succeeded in simulating a deformed composite contour based on a pair of circular ones. As a result, as before, in the analysis of individual TCs, the relationship between the intensification and dissipation of the TC system and the inflow of latent heat to the TC system from the surrounding atmosphere is shown. In the general case, the substantiation of the choice of the form and other characteristics of the enclosing contours is likely to require attracting large amounts of additional information and the development of a general concept of dynamic contours (Ermakov et al. 2017).

4.6 Concluding Remarks to this Chapter

1. This chapter demonstrates the practical application of the satellite radiothermovision approach for analyzing the evolution of tropical cyclones. A dozen tropical cyclones of different years and over various basins of the World Ocean were considered as objects of study.
2. The relationship has been shown between the change in the intensity of a tropical cyclone and the advection of latent heat from the surrounding atmosphere to the center of the cyclone. Intensification of tropical cyclone corresponds to the convergence of latent heat, and dissipation corresponds to divergence.
3. The maxima of convergent latent heat fluxes (units of Petawatts), for the first time directly calculated from satellite remote data, are one or two orders of magnitude higher than the estimates for the corresponding vertical fluxes from the ocean to the TC, and are in principle consistent with the well-known model concepts, and well explain the overall energy balance of the TC.
4. With the implemented methodology for using complex dynamical contours, the advective fluxes of latent heat, organized by a system of interacting twin typhoons, were calculated and it was shown that they determined the evolutionary nature of the considered system.
5. The joint analysis of the evolution of Hurricane Alberto (2000), the super typhoon Haiyan (2013) and the tropical storm Podul (2013) in TPW fields and composite SST fields was performed. It was shown that under conditions of an almost identical state of the ocean, the evolution scenarios of TCs can vary significantly (from rapid intensification to quasi-stationary mode and dissipation) depending on the magnitude and sign of atmospheric advection of latent heat.

References

Brand S (1970) Interaction of binary tropical cyclones of the western North Pacific Ocean. J Appl Meteorol 9(3):433–441

Dare RA, McBride JL (2011) Sea surface temperature response to tropical cyclones. Mon Weather Rev 139(12):3798–3808

Dong K, Neumann CJ (1983) On the relative motion of binary tropical cyclones. Mon Weather Rev 111(5):945–953

Dritschel DG, Waugh DW (1992) Quantification of inelastic interaction of unequal vortices in two-dimensional vortex dynamics. Phys Fluids 4A(8):1737–1744

Dworak VF (1975) Tropical cyclone intensity analysis and forecasting from satellite imagery. Mon Weather Rev 103(5):420–430

Emanuel KA (1999) The power of a hurricane: an example of reckless driving on the information superhighway. Weather 54(4):107–108

Emanuel KA (2005) Increasing destructiveness of tropical cyclones over the past 30 years. Nature 436(7051):686–688

Ermakov DM, Raev MD, Suslov AI, Sharkov EA (2007) Electronic long-standing database for the global radiothermal field of the earth in context of multy-scale investigation of the atmosphere-ocean system. Issledovanie Zemli iz kosmosa (Earth Research from Space) 1:7–13 (in Russian)

Ermakov D, Chernushich A, Sharkov E, Shramkov Ya (2011) Stream Handler system: an experience of application to investigation of global tropical cyclogenesis [electronic resource]. In: Proceedings of 34th international symposium on remote sensing of environment, Sydney, 10–15 April, 2011. http://www.isprs.org/proceedings/2011/ISRSE-34/211104015Final00456.pdf

Ermakov DM, Chernushich AP, Sharkov EA, Pokrovskaya IV (2013a) Searching for an energy source of the intensification of tropical cyclone Katrina using microwave satellite sensing data. Izvestiya Atmos Oceanic Phys 49(9):963–973

Ermakov DM, Sharkov EA, Pokrovskaya IV, Chernushich AP (2013b) Revealing the energy sources of alternating intensity regimes of the evolving Alberto tropical cyclone using microwave satellite sensing data. Izvestiya Atmos Oceanic Phys 49(9):974–985

Ermakov DM, Sharkov EA, Chernushich AP (2015) Satellite radiothermovision of atmospheric mesoscale processes: case study of tropical cyclones. In: The international archives of the photogrammetry, remote sensing and spatial information sciences—ISPRS Archives, vol XL(7/W3), pp 179–186

Ermakov DM, Sharkov EA, Chernushich AP (2016) A multisensory algorithm of satellite radiothermovision. Izvestiya Atmos Oceanic Phys 52(9):1172–1180

Ermakov DM, Sharkov EA, Chernushich AP (2017) Satellite radiothermovision analysis of the evolution of a system of interacting typhoons. Izvestiya Atmos Oceanic Phys 53(9):945–954

Ermakov DM, Sharkov EA, Chernushich AP (2019a) Evaluation of tropospheric latent heat advective fluxes over the ocean by the animated analysis of satellite radiothermal remote data. Izvestiya Atmos Oceanic Phys 55(9):1125–1132. https://doi.org/10.1134/S0001433819090160

Ermakov DM, Sharkov EA, Chernushich AP (2019b) Role of tropospheric latent heat advective fluxes in the intensification of tropical cyclones. Izvestiya Atmos Oceanic Phys 55(9):1254–1265. https://doi.org/10.1134/S0001433819090172

Ermakov DM, Raev MD, Chernushich AP, Sharkov EA (2019c) Algorithm for construction of global ocean-atmosphere radiothermal fields with high spatiotemporal sampling based on satellite microwave measurements. Izvestiya Atmos Oceanic Phys 55(9):1041–1052. https://doi.org/10.1134/S0001433819090159

Falkovich AI, Khain AP, Ginis I (1995) Motion and evolution of binary tropical cyclones in a coupled atmosphere-ocean numerical model. Mon Weather Rev 123(5):1345–1363

Ferrel W (1856) An essay on the winds and the currents of the ocean. Book and Job Printers, Cameron & Fall, p 43

Frank WM (1977) The structure and energetics of the tropical cyclone, II: dynamics and energetics. Monthly Weather Rev 105(9):1136–1150

Fritz C, Wang Z (2014) Water vapor budget in a developing tropical cyclone and its implication for tropical cyclone formation. J Atmos Sci 71(11):4321–4332

Fujiwhara S (1921) The mutual tendency towards symmetry of motion and its application as a principal in meteorology. Quarterly J Royal Meteorol Soc 47(200):287–292

Fujiwhara S (1923) On the growth and decay of vortical systems. Quarterly J Royal Meteorol Soc 49(206):75–104

Gentemann C, Smith D, Wentz F (2000) Microwave SST correlation with cyclone intensity [electronic resource]. In: Proceedings of 24th conference on hurricanes and tropical meteorology, USA, Florida, Fort Lauderdale, 22–27 May, 2000. http://images.remss.com/papers/rssconf/gentemann_ams_2000_FtLauderdale_SST.pdf

Gentemann CL, Donlon CJ, Stuart-Menteth A, Wentz FJ (2003) Diurnal signals in satellite sea surface temperature measurements. Geophys Res Lett 30(3):1140. https://doi.org/10.1029/2002GL016291

Golitsyn GS (2008) Polar lows and tropical hurricanes: their energy and sizes and a quantitative criterion for their generation. Izvestiya Atmos Oceanic Phys 44(5):537–547. https://doi.org/10.1134/s0001433808050010

Gray WM (1982) Tropical cyclone genesis and intensification. In: Bengtsson L, Lighthill J (eds) Intense atmospheric vortices. Topics in atmospheric and oceanographic sciences. Springer, Berlin, Heidelberg, pp 3–20. https://doi.org/10.1007/978-3-642-81866-0_1

Hart RE, Maue RN, Watson MC (2007) Estimating local memory of tropical cyclones through MPI anomaly evolution. Mon Weather Rev 135(12):3990–4005
Hoover EW (1961) Relative motion of hurricane pairs. Mon Weather Rev 89(7):251–255
Ivanov VN, Pudov VD (1977) Structure of the thermal wake of typhoon Tess in the ocean and estimation of the certain energy-exchange parameters under storm conditions. Typhoon-75 Gidrometidat. 1:66–82. (in Russian)
Kaplan J, DeMaria M, Knaff JA (2010) A revised tropical cyclone rapid intensification index for the Atlantic and Eastern North Pacific basins. Weather Forecast 25(1):220–241
Korolev VS, Petrichenko SA, Pudov VD (1990) Heat and moisture exchange between the ocean and atmosphere in tropical storms Tess and Skip. Sov Meteorol Hydrol (English translation) 3:92–94
Kudryavtsev V, Monzikova A, Combot C, Chapron B, Reul N, Quilfen Y (2019) A simplified model for the baroclinic and barotropic ocean response to moving tropical cyclones: 1. Satellite observations. J Geophys Res Oceans 124(5):3446–3461
Kudryavtsev V, Monzikova A, Combot C, Chapron B, Reul N (2019) A simplified model for the baroclinic and barotropic ocean response to moving tropical cyclones: 2. Model and simulations. J Geophys Res Oceans 124(5):3462–3485
Landsea CW How much energy does a hurricane release? [electronic resource]. NOAA, AOML. http://www.aoml.noaa.gov/hrd/tcfaq/D7.html
Levina GV (2018) On the path from the turbulent vortex dynamo theory to diagnosis of tropical cyclogenesis. Open J Fluid Dyn 8(1):86–114
Lin I-I, Pun I-F, Lien C-C (2014) "Category-6" supertyphoon Haiyan in global warming hiatus: Contribution from subsurface ocean warming. Geophys Res Lett 41(23):8547–8553
Makarieva AM, Gorshkov VG, Nefiodov AV, Chikunov AV, Sheil D, Nobre AD, Li B-L (2017) Fuel for cyclones: the water vapor budget of a hurricane as dependent on its movement. Atmos Res 193:216–230
Melnikov VP, Shumlsky JJ (1997) The vortex phenomena at the atmosphere. Institute of the earth's cryosphere, Tyumen, p 45. (In Russian)
Microwave OI SST Product Description. Remote Sensing Systems. [electronic resource]. http://www.remss.com/measurements/sea-surface-temperature/oisst-description/
Montgomery M, Farrell B (1993) Tropical cyclone formation. J Atmos Scences 50(2):285–310
Mori N, Kato M, Kim S, Mase H, Shibutani Y, Takemi T, Tsuboki K, Yasuda T (2014) Local amplification of storm surge by Super Typhoon Haiyan in Leyte Gulf. Geophys Res Lett 41(14):5106–5113
Palmén E, Newton CW (1969) Atmospheric circulation systems: their structural and physical interpretation, vol XVIII. Academic Press. New York, p 606
Permyakov MS (2007) Tropical cyclones: formation and development, interaction with the ocean: abstract. dis.... Dr. Phys. Sciences. Vladivostok, Russia, p 36. (in Russian)
Petty GW (1990) On the response of the special sensor microwave/imager to the marine environment—implications for atmospheric parameter retrievals. A dissertation ... for the degree of doctor of philosophy. University of Washington, Seattle, USA, p 313
Pokrovskaya IV, Sharkov EA (2006) Tropical cyclones and tropical disturbances of the world ocean: chronology and evolution. Version 3.1 (1983–2005). Poligraph Servis, p 728
Pokrovskaya IV, Sharkov EA (2016) Tropical cyclones and tropical disturbances of the world ocean: chronology and evolution. Version 5.1 (2011–2015). University Book House, p 164. ISBN 978-5-91304-661-1
Prieto R, McNoldy BD, Fulton SR, Schubert WH (2003) A classification of binary tropical cyclone-like vortex interactions. Mon Weather Rev 131(11):2656–2666
Rotunno R, Emanuel KA (1987) An air–sea interaction theory for tropical cyclones. Part II: evolutionary study using a nonhydrostatic axisymmetric numerical model. J Atmos Sci 44(3):542–561
Ruprecht E (1996) Atmospheric water vapor and cloud water: an overview. Adv Space Res 18(7):5–16
Sharkov EA (2012) Global tropical cyclogenesis. 2nd edn. Springer, Praxis, p 603

Trenberth KE, Davis CA, Fasullo J (2007) Water and energy budget of hurricanes: case studies of Ivan and Katrina. J Geophys Res 112(D23):D23106. https://doi.org/10.1029/2006JD008303

Wentz F (1997) A well-calibrated ocean algorithm for special sensor microwave/imager. J Geophys Res 102(C4):8703–8718

Wentz FJ, Hilburn KA, Smith DK (2012) Remote sensing systems DMSP SSM/I, SSMIS daily environmental suite on 0.25 deg grid, Version 7, 8 [electronic resource]. Remote Sensing Systems, Santa Rosa, CA. http://www.remss.com/missions/ssmi/

Zhao H, Tang D, Wang Y (2008) Comparison of phytoplankton blooms triggered by two typhoons with different intensities and translation speeds in the South China sea. Mar Ecol Prog Ser 365:57–65

Ziv B, Alpert P (1995) Rotation of binary cyclones—a data analysis study. J Atmos Sci 52(9):1357–1369

Chapter 5
Satellite Radiothermovision of Atmospheric Rivers

Atmospheric rivers are filamentary structures in the field of atmospheric water vapor, which provide rapid moisture transfer from the tropics to middle and high latitudes. Systematic studies of atmospheric rivers have revealed their significant role in the meridional transport of atmospheric latent heat. Starting in the intratropical convergence zone and overcoming the boundaries of the Hadley cells, atmospheric rivers often reach the polar regions and can significantly affect the climatic parameters of the latter. A particularly pronounced effect should be expected in the northern hemisphere, where there is no constant blocking effect associated with the circumpolar course. It is established that the atmospheric rivers of the Pacific basin are the cause of a significant number of extreme weather events on the west coast of North America. Atmospheric rivers over the Atlantic can have a similar effect on the regions of Western Europe.

Of significant interest is the climatology of atmospheric rivers, i.e. statistics of their characteristic parameters at climatically significant scales. Significant efforts of researchers are directed to the development and improvement of satellite data processing algorithms for automatic detection and restoration of atmospheric river parameters. This chapter discusses the unique capabilities of satellite radiothermovision to ensure further progress in this area of research. Section 5.1 summarizes the problems of studying atmospheric rivers from satellite radiometry data and ways to overcome them using satellite radiothermovision. Section 5.2 discusses a new algorithm for the automatic detection of atmospheric rivers based on interpolated fields of the total precipitable water. Section 5.3 demonstrates the capabilities of satellite radiothermovision for calculating latent heat fluxes generated in atmospheric rivers and comparing them with "background" fluxes outside atmospheric rivers. The presented results were obtained and published mainly in the works (Ermakov 2017, 2019; Ermakov and Chernushich 2018).

D. M. Ermakov, *Satellite Radiothermovision of Atmospheric Processes*, Springer Praxis Books,
https://doi.org/10.1007/978-3-030-57085-9_5

5.1 Problems of Detecting Atmospheric Rivers

As noted in Chap. 2, the concept of "atmospheric rivers" is introduced informally in the context of the analysis of a number of remote observation data (Newell et al. 1992; Zhu and Newell 1994, 1998) and raises some objections (Bao et al. 2006; Knippertz and Wernli 2010; Gimeno et al. 2014). In this regard, convincing proof of some of the concepts formulated to date (for example, on the decisive role of atmospheric rivers in the meridional transport of latent heat at mid-latitudes, (Zhu and Newell 1998)) requires a deep systematic approach. Its purpose should be, as far as possible, an exhaustive analysis of the "filamentary" structure of the atmospheric circulation of latent heat, which would make it possible to reliably detect atmospheric rivers in the fields of geophysical parameters (primarily the total precipitable water, TPW) of the lower troposphere and calculate the energy and mass transfer characteristics associated with them. Such an approach would allow one of the central problematic questions to be answered: are all "filamentary" formations in the field of TPW characterized by certain size ratios and geographical location "rivers", i.e. provide latent heat fluxes that significantly exceed the "background" values of fluxes outside these rivers in power.

Currently, such a systematic approach is developing most consistently with respect to the atmospheric rivers of the northern hemisphere and, first of all, for the northeast Pacific (Wick et al. 2013; Gimeno et al. 2014). An important incentive for these studies is the need to study and assess the risks of extreme weather events and natural disasters periodically caused by atmospheric rivers on the west coast of North America. It should also be noted interesting parallels with the task of a systematic study of high-altitude jet flows (Nerushev et al. 2018, 2019). Despite the fundamental differences between the objects of study and the input information (Nerushev and Kramchaninova 2011), the tasks are close in a number of methodological aspects (analysis of the dynamics of geophysical fields reconstructed from satellite data). High-altitude jet streams, like atmospheric rivers, can pose a significant danger (in particular, for aviation). Statistics of their intensity and frequency of formation, as in the case of atmospheric rivers, can be considered as an important indicator of regional and global climate changes (Nerushev et al. 2019).

Approaches to solving the problem of a systematic study of atmospheric rivers are developing in two main directions: analysis of the spatial structure of the TPW field according to remote sensing data (Ralph et al. 2004; Matrosov 2013), modeling or reanalysis (Dettinger et al. 2011) and analysis of vertically integrated advective water vapor flows using numerical models (Zhu and Newell 1994, 1998). At the same time, an integrated approach that combines the use of remote data and model estimates is becoming more common (Wick et al. 2013). However, it is noted that the developed approaches using satellite data are not characterized by universal applicability for the detection and analysis of atmospheric rivers on a global scale, over all ocean basins. In (Wick et al. 2013), the main problems of detecting atmospheric rivers (AR) according to satellite radiometry remote data are listed:

(1) data gaps in some cases significantly complicate the automatic detection of AR;
(2) the numerical criteria for the detection of ARs, developed on a limited scope of observations over individual basins of the World Ocean, require verification and refinement for universal use on a global scale;
(3) to improve the quality of detection of AR, it is desirable to have synchronous estimates of the TPW fields and advection of latent heat.

The rest of this section shows how these difficulties can be overcome with the satellite radiothermovision approach.

5.1.1 Data Gaps

The TPW fields reconstructed from satellite radiometry data contain gaps (see Chap. 3). Part of the omissions is due to the technical features of the survey or malfunctioning devices. In addition, there are systematic omissions (lacunae) associated with the divergence of the satellite swaths at low and middle latitudes, Fig. 5.1a.

Automatic detection of AR involves the search for narrow extended atmospheric features, more than 2000 km in length and less than 1000 km (Wick et al. 2013) or 500 km in width (Gimeno et al. 2014), characterized by high TPW values. Gaps in the analyzed fields lead to fragmentation of search objects, significantly complicating the detection of its parts and the assessment of its full linear dimensions. Satellite radiothermovision, due to the algorithm for lacunae stapling and spatiotemporal interpolation, provides restoration of the TPW fields without gaps and dramatically facilitates the automatic detection of extended atmospheric features, Fig. 5.1b.

In Fig. 5.1, AR is clearly visible, starting in the intertropical convergence zone of about 20° N, 170° E and stretching northeast to the west coast of North America. In

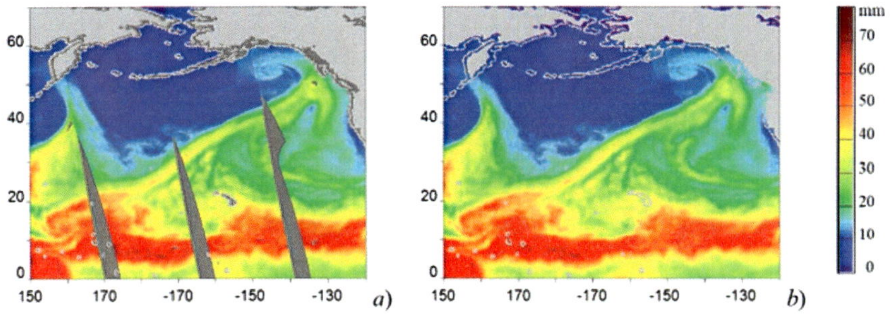

Fig. 5.1 A fragment of the TPW field over the Pacific Ocean (color scale in mm on the right) for December 1, 2016: **a** mosaic according to SSMIS F16 and SSMIS F17; **b** a product of satellite radiothermovision. At the edges, the geographical coordinates of the fragment in degrees; positive values for the northern and eastern hemispheres, negative values for the southern and western

the standard mosaic of satellite remote data, the AR splits into 3 fragments separated by gaps (shown in dark gray in Fig. 5.1a).

Since the proposed AR detection algorithms (Wick et al. 2013) do not imply additional data interpolation (except for the simplest median filtering), the formation of a complete AR image based on the available SSMIS data is impossible. Satellite radiothermovision allows to restore a holistic picture of the phenomenon, Fig. 5.1b.

5.1.2 Setting Detection Criteria

As noted in (Wick et al. 2013), the criteria for detecting ARs related to a particular basin of the World Ocean may require refinement to adapt to other areas of observation. The development of universal criteria for automatic detection requires the processing and analysis of a large amount of data on a global observation scale. A successful solution to this problem is associated with overcoming a number of technical difficulties: ensuring the rapid downloading (or construction) of the TPW fields of the global coverage; the implementation of interactive control of processing procedures in order to optimally configure the detection criterion; effective dynamic visualization of data and processing results.

All these aspects of the problem find solutions in the approach of satellite radiothermovision, which were implemented during the development of the geoportal of satellite radiothermovision and of the ICAR project (see Chap. 7). The geoportal currently contains an array of global fields of TPW on a regular grid of 0.25° in the continuous observation range from 2003 to 2019, interpolated in 3 h increments. The functionality of the ICAR project provides the ability to describe various processing algorithms for these data, remote execution of these algorithms on the geoportal server and visualizing the processing results. Figure 5.2 illustrates the application of the simplest threshold criterion for identifying areas of high TPW values.

Areas of TPW values exceeding thresholds of 20, 23, 26, and 30 mm introduced by analogy with (Wick et al. 2013) are shown. In Fig. 5.2a, three ARs are clearly visible. The first crosses the northeastern Pacific Ocean, reaching the coast of North America about 35° N. The second begins in the east of the Pacific Ocean, crosses Mexico and the Gulf of Mexico west of the Yucatan, passes over the eastern coast of North America above Florida and stretches through the center of the North Atlantic along 40° N. The third crosses the Atlantic in a predominantly meridional direction and reaches northern Scotland about 55° N. It should be noted that two Atlantic ARs can be skipped during automatic detection due to their width: scale sections of 1000 km in Fig. 5.2a fit entirely inside the AR images in their widest places.

The increase in the threshold to 23 mm (Fig. 5.2b) has almost no effect on the image of the Pacific AR, while the width of the other two is significantly reduced. In this case, however, gaps arise in the image of the eastern AR, in particular, indicated by the arrow in Fig. 5.2b at about 50° N, 20° W, which can lead to noticeable errors in determining the length of the AR. A further increase in the threshold to 26 mm (Fig. 5.2c) causes the destruction of the northern part of the image of the Pacific AR,

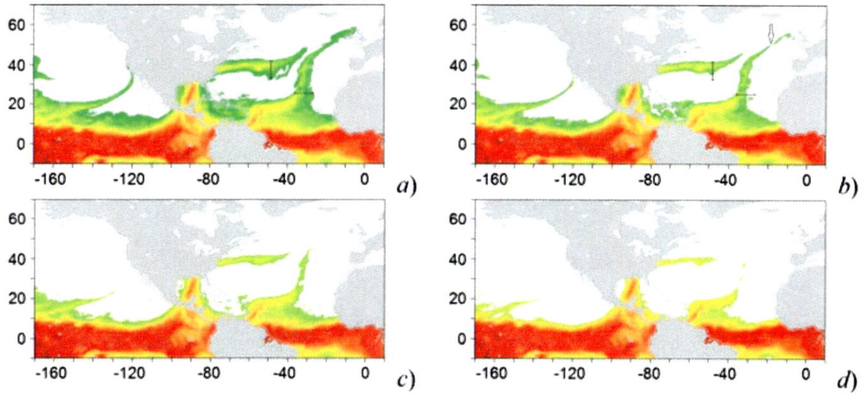

Fig. 5.2 The TPW field over the northeast Pacific Ocean and the North Atlantic for 03/11/2016 using the satellite radiothermovision algorithm (color scale and coordinate notation—as in Fig. 5.1). TPW values above the following thresholds are shown: **a** 20 mm; **b** 23 mm; **c** 26 mm; **d** 30 mm. The scale sections in the upper figures correspond to 1000 km

which now ends at about 20° N. Images of Atlantic ARs reach approximately 40° N. Finally, at a threshold of 30 mm, the image of the easternmost AR itself is destroyed, also about 20° N. Changes in the length of the central AR are still relatively small.

Attention is drawn to the instability of the linear dimensions of the AR images with respect to small changes in the detection threshold. More promising are multi-threshold and/or adaptive algorithms for detecting ARs. Here, the approach of satellite radiothermovision can be useful for the effective organization of distributed processing and analysis of remote data. The functionality of the ICAR project allows for unified description and testing of various processing algorithms. The geoportal of satellite radiothermovision provides easy quick access to the array of global TPW fields (and other geophysical parameters) according to long-term observations.

5.1.3 Advection Field Accounting

Combining the TPW field with the vector fields of advection of water vapor can significantly improve the detection results, but so far this possibility has been considered only in connection with the use of synoptic numerical simulation schemes (Wick et al. 2013). Satellite radiothermovision offers an alternative approach, providing the calculation of the advection field directly from a series of satellite observations, Fig. 5.3.

Figure 5.3 illustrates the directions and relative velocities of advection observed in the TPW fields. To simplify the picture, the advection field is sampled out with a step of 2°, the vectors are represented by non-directional segments in the same projection as the TPW field. The geometric correction and normalization of the

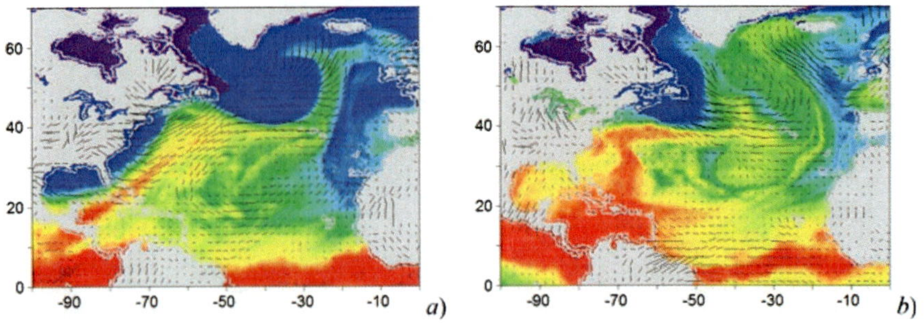

Fig. 5.3 The TPW field over the North Atlantic with superimposed elements of the advection field; calculation on the dates of observations: **a** 03/21/2016; **b** 06/01/2016. The color scale and designations of coordinates—as in Fig. 5.1

elements of the advection field for the transition to the absolute values of the velocity and calculation of the latent heat fluxes are discussed in detail in (Ermakov et al. 2017, 2019). It is important to note here that the direction of advection within the image of the AR, generally speaking, is not parallel to its axis due to shifts and evolution of the AR during interaction with other atmospheric systems, for example, extratropical cyclones. This circumstance must be taken into account in the algorithms for detecting and recovering AR parameters. Since the average TPW field gradient along the AR axis is much smaller than across the axis, to expand the set of potential tracers when analyzing movement inside the AR, it is advisable to carry out calculations on a more detailed grid (Ermakov et al. 2015, 2016) and/or attract additional data (for example, the fields of cloud liquid water, the precipitation rate).

5.1.4 Satellite Data Synchronization

The traditional mosaic approach to constructing fields of geophysical parameters has another drawback in the context of the considered climatology problem of the AR. ARs are extended objects that can span several time zones at once. Measurement data from solar-synchronous satellites belong to the same local time and, accordingly, the intervals between measurements of different fragments of the AR can be several hours. With the rapid evolution of ARs, such asynchrony may prove essential for the remote diagnostics of their parameters. It should also be noted that in the zonal relation, the spread of AR occurs in the east direction, and satellite monitoring from solar-synchronous orbits is carried out in the opposite, to the west.

Satellite radiothermovision provides bringing the interpolated TPW fields to a single point in universal time (Ermakov et al. 2016) to reconstruct the "instant" picture of atmospheric processes, Fig. 5.4.

In Fig. 5.4a, the "instantaneous" TPW field is constructed, which closely corresponds to the time instant of the passing of SSMIS F16 and SSMIS F17 over 170° E,

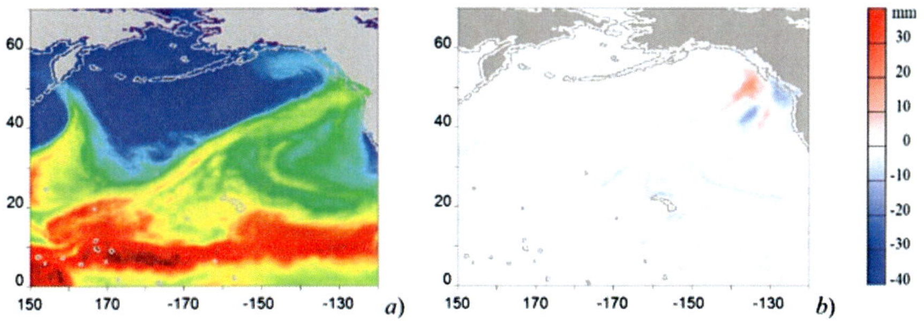

Fig. 5.4 The TPW field (color scale—see Fig. 5.1), recalculated to 12/02/2016 01:00 UTC (**a**), and the difference of the fields shown in Figs. 5.1b and 5.4a (**b**), the color scale—to the right

i.e. in the area of the "source" of the AR, shown in Fig. 5.1. It should be noted that the image of the AR in Fig. 5.1 is built with a fixed local time. This means that the end of the AR, reaching the coast of North America at about 130° W (i.e., 60° east of the beginning of the AR), was observed 4 h earlier than its beginning. During these 4 h, there was a change in the axis of the AR and a significant shift in the place of its landfall in a southeast direction. This is clearly seen in Fig. 5.4b, which illustrates the difference in the fields shown in Figs. 5.1b and 5.4a, respectively. As expected, minimal differences are observed in the zone of 165–175° E. The most noticeable differences are associated with the rapid evolution of AR. A positive difference (red tones) corresponds to the AR image in Fig. 5.1b (in the local time format), and a negative difference (blue tones) corresponds to the AR image in Fig. 5.4a (in the universal time format).

Thus, satellite radiothermovision algorithms are very relevant for the study and refinement of the characteristics of ARs, which feature is a combination of significant zonal extent and rapid evolution.

5.2 Synthesis of an Algorithm for Automatic Detection of Atmospheric Rivers

For the above reasons for the ambiguity of the term "atmospheric rivers", the approach to the analysis of the fine structure of atmospheric circulation is based on a formal search of all elements (objects) in the TPW field of the atmosphere that satisfy the following selection criteria:

(1) significant length (>2000 km);
(2) small width (<1000/500 km);
(3) a high gradient of TPW values at the boundaries of objects, especially in mid-latitudes;
(4) latitudinal coverage from the boundaries of the intertropical convergence zone (ICZ) to high latitudes;

(5) simply connected and lack of cycles;
(6) the existence time is several days.

The main goal is the detection and exhaustive listing of all processes observed in the TPW field that are potentially ARs, and the exact localization of the axes of these ARs. Successful implementation of this step provides a further systematic approach to analysis:

(1) the formation and evolution of AR;
(2) additional characteristics of AR in the vicinity of the main axis (precipitation intensity, cross-sectional width, geophysical field gradients, etc.)
(3) the restored values of the velocity of advection and the power of the latent heat flow associated with AR.

This, in turn, opens up the possibility of studying the role of AR in the meridional circulation of atmospheric latent heat, as well as proving (or refuting) a number of theses regarding AR, which are currently a hypothetical generalization of the results of some specific studies.

The proposed and described below scheme for the automatic detection of AR and AR-like phenomena in the TPW field is illustrated in general form in Fig. 5.5.

The left side of Fig. 5.5 lists the main algorithmic steps of the flowchart. On the right side, a specific processing example illustrates the steps of data conversion, which ultimately allow the axes of potential (automatically detected in accordance with the above criteria) ARs to be placed on the TPW field. The processing steps are discussed below in more detail.

5.2.1 *Specification of the Boundaries of the Basin*

Since, as indicated above, the average AR characteristics may depend on the geographical location (differ for different basins of the World Ocean), it makes sense to perform AR detection in the TPW field over different basins separately. For this purpose, it was proposed to identify five large basins: the northern and southern Pacific Ocean, the northern and southern Atlantic and the Indian Ocean. In this case, for an unambiguous definition of the basin, it is enough to set its western and eastern borders (for some fixed meridians), as well as select the southern or northern hemisphere. Since data processing is carried out on a regular coordinate grid, setting the western and eastern boundaries of the water area corresponds to setting the minimum and maximum index of the grid columns (horizontal coordinates of cells) $i_{\min}, i_{\max}$. As an example, Fig. 5.6a shows a fragment of the TPW field corresponding to the North Pacific defined in this way.

At high latitudes, it is proposed to fix the boundary of the water area at a parallel of 70°. At low latitudes, a finer approach to localizing the latitudinal boundary of the basin is required, since the problem arises of separating the region of higher TPW values associated with AR from the ICZ region (Wick et al. 2013). For this purpose,

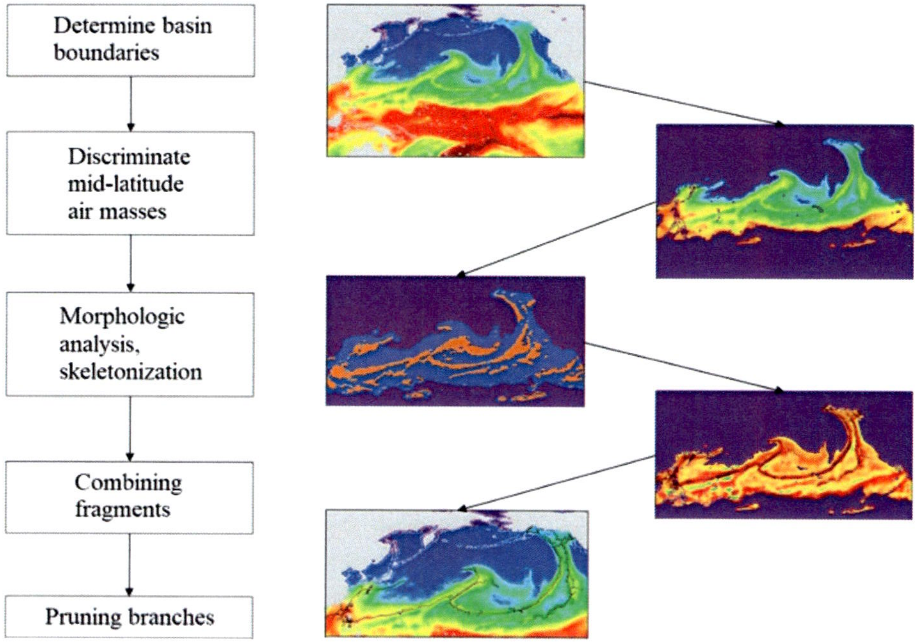

Fig. 5.5 Schematic flowchart of automatic detection of atmospheric rivers in the total precipitable water field of the atmosphere

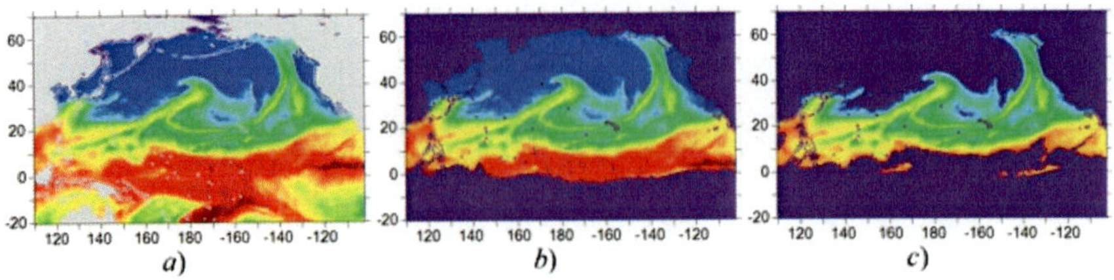

Fig. 5.6 **a** Selection of a fragment of the TPW field for a given basin (Northern Hemisphere, Pacific Ocean), the color scale of the TPW values and designation of coordinates—as in Fig. 5.1; **b** refinement of the latitudinal border in the tropics (the mask of the areas excluded from the analysis is shown in purple); **c** the allocation of air masses of mid-latitudes (the mask of the areas excluded from the analysis is shown in purple)

the "axis" of the ICZ is first localized. For a fixed value of longitude (i.e., a fixed value of the column index i, $i_{\min} \leq i \leq i_{\max}$), the row index j_0 (the vertical coordinate of the cell) is calculated according to the following rule:

$$j_0 = \sum_{j=j_{20S}}^{j_{20N}} j \cdot W(i, j) \Bigg/ \sum_{j=j_{20S}}^{j_{20N}} W(i, j), \qquad (5.1)$$

where j is the column number running through all the values between the indices corresponding to 20° south latitude and 20° north latitude; W is the value of the TPW field in the node (i, j). Thus, j_0 is the position of the "center of mass" of the TPW values at a given longitude within the latitudinal belt of ±20°. In contrast to the formal choice of the lower boundary of the water area at the equator, the proposed method allows to take into account the seasonal migration of ICZ, as well as its local and short-term deformations, Fig. 5.6b.

5.2.2 *Discrimination of Air Masses of Middle Latitudes*

One of the most specific characteristics of AR is a significant excess of the TPW values inside it over the "background" values, which is especially pronounced outside the tropics. At the same time, however, the detection of such an excess using the simplest threshold filtering with absolute limit values encounters significant difficulties (Wick et al. 2013), and an adaptive technique for finding optimal threshold values is required.

In the scheme presented in the book, an important role is played by the allocation of air masses of middle latitudes on the analyzed fragment of the TPW field. It should be noted that, in general, the classification of air masses by type and the terminology used in this connection are not well established and their various options can be applied in the context of different tasks. So, the authors (Palmén and Newton 1969,

p. 103) directly indicate: "By considering the large-scale advective processes in the atmosphere and the modifications of the different air properties due to radiation, exchange of heat and moisture between the earth's surface and the atmosphere, condensation processes, etc., the subdivision of air masses can be carried very far". And further: "For our purpose, it therefore seems convenient to divide the hemispheric air masses into only three principal classes: *tropical air (TA), middle-latitude air (MLA), and polar air (PA)*". In the case considered in this book, it was also convenient to consider three types of air masses over the ocean: Arctic (Antarctic), tropical, and mid-latitudes. If we again refer to Fig. 5.6, then in the selected color scheme, the former are displayed in blue, the latter in red, and the intermediate between them (mid-latitudes) are yellow–green. Such a conditional classification turns out to be quite acceptable in this case, since, in fact, the detection problem under consideration is very close to a wide class of technical vision problems, initially posed in a weakly formalized subject area.

To formalize the problem of dividing air masses into the indicated types, a preliminary hypothesis was adopted that the air masses of each of the three types are characterized by the normal distribution law of the TPW values relative to a certain average, which is a function of the type of air masses and varies within certain limits depending on the basin, season, and (possibly) other factors. This gives rise to the general idea of classifying the selected fragment of the TPW field (Fig. 5.6b) by type of air mass. For this, it is necessary to construct a histogram of the TPW values over the entire fragment, smooth it to reduce sharp emissions (associated with both the fractional noise effect when rounding the TPW values to average and with limited sample sizes) and approximate the resulting smooth curve by the sum of several Gaussian functions.

A nontrivial algorithm is the approximation of a smoothed histogram by the sum of Gaussian functions. As a first approach it was implemented iteratively. Let's denote $h(w)$ the number of cells included in the fragment of the TPW field under consideration, the value of the TPW in which equals w. Thus, $h(w)$ is the initial histogram of a fragment of the TPW field. The result of its smoothing is denoted $\hat{h}(w)$. It was found that satisfactory smoothing results are obtained with averaging of $h(w)$ values in a sliding window 7 units wide (i.e. 7 kg/m^2), Fig. 5.7.

On the smoothed histogram the position w_1 of the first (closest to 0) local maximum is located according to the following rule:

$$\hat{h}(w_1) \geq \hat{h}(w), \quad w = 0 \ldots 15. \tag{5.2}$$

The limit value of the search range of 15 kg/m^2 was established empirically based on the analysis of a large amount of data. Then, the optimal parameters and a Gaussian function are selected that approximate the slope closest to the edge and in the vicinity of the local maximum found:

$$\hat{h}(w) \approx g_1(w) \equiv A_1 e^{-\frac{(w-w_1)^2}{\sigma_1^2}}, \quad 0 \leq w \leq 1.5w_1. \tag{5.3}$$

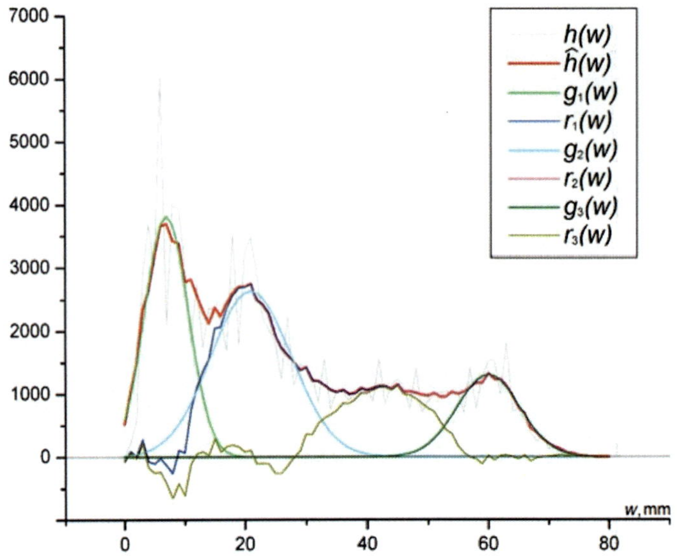

Fig. 5.7 Analysis of the histogram of TPW values of the selected field fragment (see Fig. 5.6b). On a horizontal scale—TPW values, on a vertical—the number of corresponding TPW values. Thin gray line is the initial histogram; the red line is a smoothed histogram; light green line—approximation of the first peak; the blue line is the residue from the subtraction from the smoothed histogram of the first approximated peak; cyan line—approximation of the second peak; the pink line is the residue from the subtraction of the second approximated peak; dark green line—approximation of the fourth peak; the yellow–green line is the remainder of the subtraction of the fourth peak. Designations correspond to Eqs. (5.2–5.11). See explanations in the text

To this end, the problem is reduced to a linear form by logarithm

$$\ln\left(\hat{h}(w)\right) \approx \ln A_1 - \frac{(w - w_1)^2}{\sigma_1^2} \tag{5.4}$$

and is solved by the classical least-squares method with respect to unknowns $\ln A_1$ and σ_1^{-2} with a fixed w_1, Fig. 5.7.

Next, the residual part of the smoothed histogram is calculated in the form

$$r_1(w) = \hat{h}(w) - g_1(w), \tag{5.5}$$

for which a new (closest to zero) local maximum w_2 is searched for in the interval $[2w_1, 40]$, where $w_1 \leq 15$ is the position of the local maximum found earlier by (5.2):

$$r_1(w_2) \geq r_1(w), \quad w = 2w_1 \ldots 40. \tag{5.6}$$

As before, the second boundary value of 40 kg/m^2 was established empirically. Next, the approximation problem is again solved:

$$r_1(w) \approx g_2(w) \equiv A_2 e^{-\frac{(w-w_2)^2}{\sigma_2^2}}, \quad (w_1 + w_2)/2 \le w \le w_2 + (w_2 - w_1)/2. \quad (5.7)$$

The obtained values A_2 and σ_2^2 are used to calculate the new residue:

$$r_2(w) = r_1(w) - g_2(w). \quad (5.8)$$

Finally, a search is made for the local maximum of the residue $r_2(w)$, w_3, which is closest to the maximal w values:

$$r_2(w_3) \ge r_2(w), \quad w > 40. \quad (5.9)$$

Next, the approximation problem is solved

$$r_2(w) \approx g_3(w) \equiv A_3 e^{-\frac{(w-w_3)^2}{\sigma_3^2}}, \quad w \ge w_3 - 5, \quad (5.10)$$

and the obtained values A_3 and σ_3^2 are used to calculate the residue

$$r_3(w) = r_2(w) - g_3(w). \quad (5.11)$$

As a result, the range of w values corresponding to the air masses of middle latitudes is determined on the considered fragment of the TPW field. The lower value, $w_{\min}$, is determined by the intersection of $g_1(w)$ and $g_2(w)$. The upper value, $w_{\max}$, is determined by the intersection of $g_3(w)$ and $r_3(w)$. A manifold of cells containing TPW values $w_{\min} \le w \le w_{\max}$ is distinguished, and further analysis is carried out only of the selected cells (it is assumed that they conditionally correspond to air masses of middle latitudes). Indeed, the above algorithm successfully localizes the region in which the potential atmospheric river is located, Fig. 5.6c. In this case, two important aspects should be noted. The first is that the cells selected in this way often do not form a simply connected region, and some selected fragments cannot be geographically assigned to middle latitudes. These fragments are automatically excluded from analysis during the execution of further steps of the algorithm.

The second aspect is of great interest and consists in the fact that the residue $r_3(w)$, contrary to the initial assumption, in most of the cases considered is not equal to zero, but can be satisfactorily approximated by another Gaussian function, Fig. 5.7. This fact was proved with a more advanced version of the algorithm which searched the best solution of the approximation problem in twelve-dimensional space ($A_1, A_2, A_3, A_4, w_1, w_2, w_3, w_4, \sigma_1, \sigma_2, \sigma_3, \sigma_4$) using Levenberg-Marquardt and genetic algorithms (Ermakov and Chernushich 2018). Hence, it can be suggested that the air masses of middle latitudes (understood in the above sense) from the point of view of the distribution of TPW values are a two-phase mixture that practically does not come to a uniform state under the disturbing influence of the tropical and Arctic/Antarctic air masses bordering them. However, as already noted, a violation of the original hypothesis does not affect the possibility of efficient localization of the cells (and the corresponding narrowing of the range of TPW values) related to the manifestation of AR in the TPW field. This allows to successfully implement the following steps of automatic analysis.

5.2.3 Morphological Analysis

In the framework of the considered problem of automatic detection of ARs, the morphological analysis, in the narrow sense, was understood as the analysis of the region of the TPW field selected at the previous step, aimed at searching and localizing fragments of AR images in the TPW field. According to the search conditions, these fragments should be characterized by significant linear dimensions in one dimension, relatively high values of the TPW and the gradient of the TPW field in the transverse direction. Since selection criteria based on fixed threshold values turn out to be not sufficiently effective (see above), the algorithm, in fact, implements an adaptive approach that takes into account the "environment" of the search object, i.e. state of the TPW field in the vicinity of the proposed axis of the AR. This, to a large extent, but not completely, solves the problems of "false alarm" (a too soft selection criterion that highlights a lot of false fragments) and "missed targets" (a too hard criterion that leads to excessive fragmentation of search objects). The solution is refined at the next steps of the algorithm, which, thus (to a large extent, arbitrarily), are outside the scope of the morphological analysis described here.

The analysis procedure consists of two algorithmically similar parts: the search for longitudinally extended fragments of the AR and the search for latitudinally extended fragments of the AR. As noted above, ARs have a predominantly longitudinal orientation. For this reason, the significance of longitude and latitude gradients of the TPW field during automatic detection of ARs is different. Priority should mainly be given to the search for longitudinally oriented fragments. However, the search for latitudinally-oriented ones is necessary for the restoration of partially fragmented images of the AR. A possible approach is to analyze the full gradient of the TPW field with adaptive weights that regulate the relative importance of derivatives in orthogonal directions. However, an alternative approach seems to be simpler and ideologically clear, consisting in an independent search and localization of fragments with a longitude and latitudinal orientation with further additional selection and combining of results.

The algorithm for searching for longitudinally oriented fragments is described in detail below. Its block diagram is shown in Fig. 5.8. The algorithm consists of the following main steps:

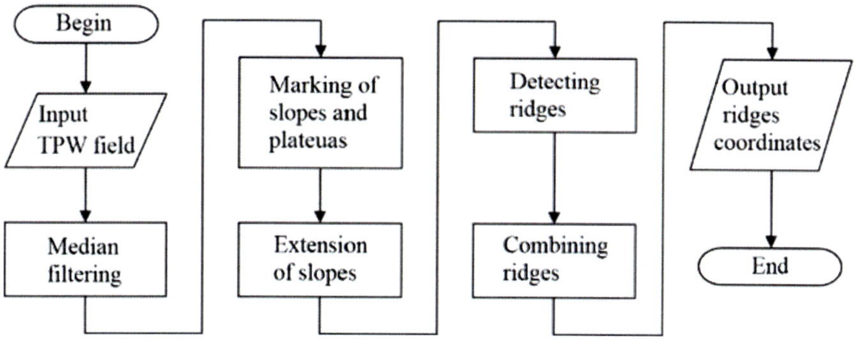

Fig. 5.8 Block diagram of the morphological analysis of a fragment of the TPW field

(1) Median filtering (smoothing the TPW field in a sliding window of 7 × 7 cells);
(2) Marking of slopes and plateaus (according to the sign of the derivative in the latitudinal direction);
(3) Extension of slopes;
(4) Localization of ridges (cells in which the direction of the slope changes);
(5) Combination of ridges.

The **median filtering** step was used to smooth the initial TPW field, since the magnitude and sign of the derivative field are sensitive to outliers, which can be associated with individual errors in the reconstruction of the TPW values as well as with sampling noise. Filtering is implemented in a sliding square window of 7 × 7 cells. The initial 49 values of the TPW field in the cells covered by the window are sorted in ascending order, the central value in the resulting sequence is assigned to the cell of the smoothed field, relative to which the window is centered:

$$\widetilde{W}(i, j) = median\{W(m, n)\}_{m=i-3,n=j-3}^{m=i+3,n=j+3}. \tag{5.12}$$

Here, $\widetilde{W}(i, j)$ is the calculated value of the smoothed TPW field in the cell (i, j), $W(m, n)$ is the value of the initial TPW field in the cell (m, n), the expression in curly brackets means the set of values $W(m, n)$ in cells that are no more than three cells in each dimension from the central one (i, j), $median$ is the operation of calculating the median value. It is accepted that an increase in the i index corresponds to the eastward direction, an increase in the j index corresponds to the northward direction.

The procedure is repeated for all cells (i, j) distant from the edge of the analyzed area by at least three cells in both directions. The smoothing result is illustrated by the example in Fig. 5.9.

The step of **marking the slopes and plateaus** consists in assigning to each cell an integer index of the "local slope" $S(i, j)$ of the derivative of the smoothed TPW field along the meridian:

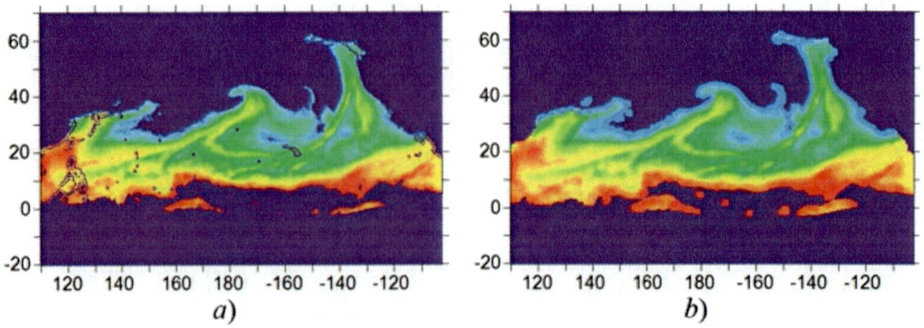

Fig. 5.9 Smoothing the TPW field by median filtering: **a** a fragment of the initial TPW field; **b** the same fragment after smoothing. The color scale of the TPW values and the notation of coordinates—as in Fig. 5.1

$$S(i,j)=\begin{cases}1, if\ \widetilde{W}(i,j+1)>\widetilde{W}(i,j),\\ 0, if\ \widetilde{W}(i,j+1)=\widetilde{W}(i,j),\\ -1, if\ \widetilde{W}(i,j+1)<\widetilde{W}(i,j).\end{cases} \tag{5.13}$$

The goal of this step is to find the set of cells where local maxima of the smoothed TPW field arise, localizing fragments of the AR axis as boundaries between the "slopes" facing north and south, respectively, Fig. 5.10.

However, the smoothing performed in the previous step leads to the fact that the field near this boundary forms a "plateau" (the region of equal TPW values). To obtain a border in the form of a linear object (with a transverse dimension of one or two grid cells), the procedure for **extending the slopes** is additionally performed.

The procedure for **extending slopes** is performed iteratively until iterations change at least one $S(i, j)$ zero value to nonzero according to the following rule. If among the values in the neighboring cells $S(i-1, j)$, $S(i+1, j)$, $S(i, j-1)$, $S(i, j+1)$ there is at least one nonzero, then the new $S'(i, j)$ value is taken equal to 1, if among the above positives ones prevail negative ones, otherwise it is taken equal

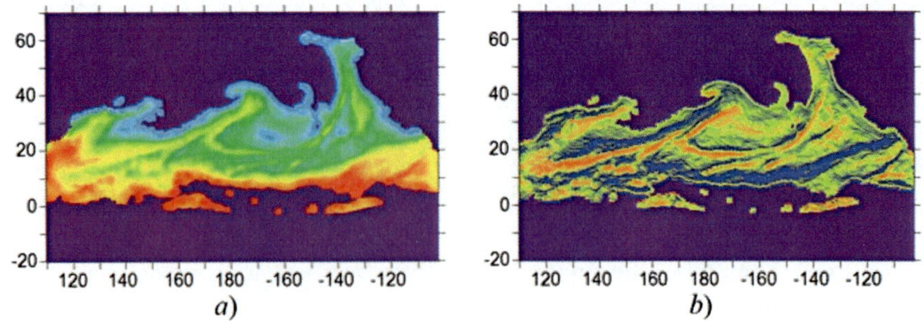

Fig. 5.10 Marking the slopes and plateau of the smoothed TPW field: **a** a fragment of the smoothed TPW field (as in Fig. 5.9b); **b** a map of slopes and plateaus: the northern slopes are marked in blue, the southern slopes are orange, the plateau is yellow–green, see the text for explanations

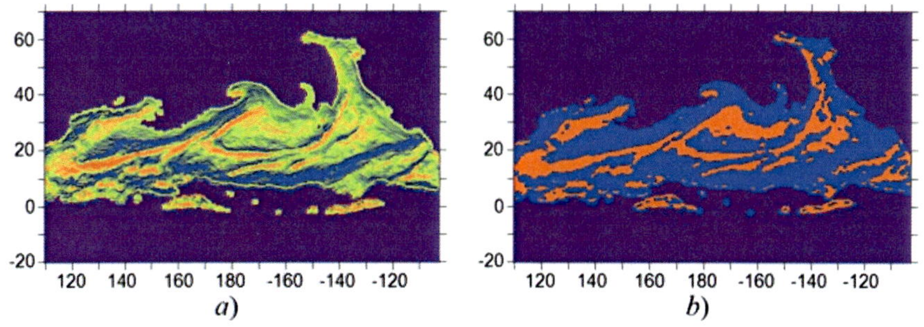

Fig. 5.11 The procedure for extending slopes: **a** the initial map of marked slopes and plateaus (as in Fig. 5.10b); **b** the result of the slope extension procedure, see the explanations in the text

to -1, except if the slope index values in all neighboring cells are zero, in which case $S'(i, j)$ is assumed to be 0. The meaning of the described procedure is that the direction of the slope is attributed to the plateau cell adjacent to this slope.

After applying these rules to all cells (i, j) (except the boundary cells of the analyzed area), the field $S(i, j)$ is replaced by $S'(i, j)$. If at least one difference occurs in the indicated fields, the next iteration is performed. As a result, the slope index in all cells of the analyzed region acquires a nonzero value. An illustrative example is shown in Fig. 5.11.

Next, the procedure for **localization of ridges** (fragments of the axes of the potential AR) is carried out. The cells located on the border of the slopes with different signs are marked as belonging to the axis of the potential AR, for which a mask is formed:

$$A(i, j) = \begin{cases} 1, \; if \; S(i, j) = 1 \; and \; S(i, j+1) = -1, \\ 0, \; otherwise. \end{cases} \tag{5.14}$$

The result of the procedure is illustrated by an example in Fig. 5.12.

Fig. 5.12 Localization of ridges: cells corresponding to ridges according to criterion (5.14) are marked in white in the figure; the background is the same as in Fig. 5.11b (on an enlarged scale)

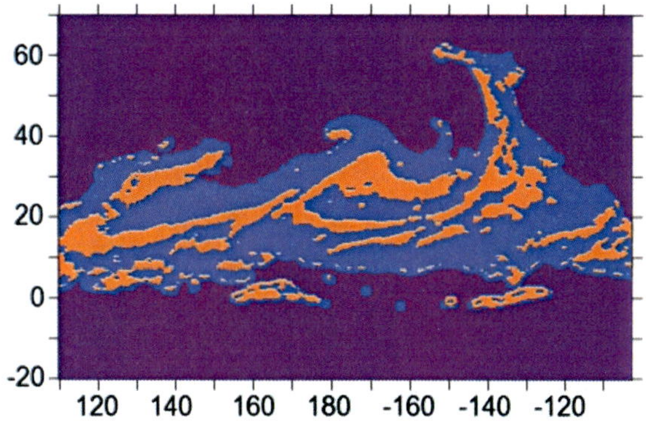

As can be seen from Fig. 5.12, the direction of the sign change of the slope is important, because depending on it, the boundary between the slopes can correspond to either a local maximum of the TPW field or a local minimum. The search for local maxima is of interest, but the configuration of the AR can be complicated. If the axis of the AR forms a meander-like figure, then part of the axis passes near the "hollow" region (the fragment highlighted by a rectangular frame in Fig. 5.12). The $A(i, j)$ values in these cells are equal to zero, and the detected image of the axis of the AR undergoes a gap. The step of **combining the ridges** is aimed at eliminating such gaps.

To implement the step of **combing the ridges**, the concept of an AR *fragment* is formalized and the concept of a boundary cell is introduced. A chain of length 1 containing a cell (i, j) belonging to the ridge $A(i, j) = 1$ is the totality of this cell and all its neighbors, including the diagonal ones, which are assigned to the axis of the AR, in other words, the totality of cells:

$$\{m, n\} : A(m, n) = 1, i - 1 \leq m \leq i + 1, j - 1 \leq n \leq j + 1, A(i, j) = 1. \tag{5.15}$$

The chain of length k containing a cell (i, j), $C^k(i, j)$ is the set of all cells of a chain of length $k-1$, $C^{k-1}(i, j)$ containing the cell (i, j), and all the cells (m, n), $A(m, n) =$ 1 not included in the chain $C^{k-1}(i, j)$, but neighboring (including diagonally) at least one cell of the chain $C^{k-1}(i, j)$.

A *fragment* $C(i, j)$ of an AR containing a cell (i, j) is a chain $C^k(i, j)$ of maximum length k (i.e., including all other chains containing a cell (i, j) as its subsets). Obviously, if two fragments $C_1(i, j)$ and $C_2(i, j)$ contain a common cell (i, j), then they are identical. Non-identical fragments do not contain common cells and, therefore, do not interconnect.

A cell (i, j) is called "boundary" if $A(i, j) = 0$ (i.e., the cell is not assigned to the axis of the AR), $S(i, j) = 1$ and the slope index value in at least one of the neighboring cells is -1, i.e. the cell is located on the border of the slopes of different signs. In addition, the concept of the order of the boundary cell is introduced. A boundary cell (i, j) is of the order of 1 with respect to the fragment $C(m, n)$ if one of its neighbors, including the diagonal ones, belongs to $C(m, n)$. A boundary cell has an order of k with respect to the ridge $C(m, n)$ if one of its neighbors, including diagonal ones, is a boundary cell of order of $k - 1$ with respect to the fragment $C(m, n)$.

The combination of two (non-identical) fragments C_1 and C_2 is carried out according to the following rule. If some cell (x, y) is a boundary of order 1 with respect to the fragment C_1 and simultaneously boundary of order $k \leq k_{\max}$ (where $k_{\max}$ is a fixed number) with respect to the fragment C_2, then a new fragment C is constructed that contains all cells of the fragments C_1 and C_2, as well as a sequence of boundary cells of the orders $k, k - 1, \ldots, 1$ with respect to the ridge C_2, starting with (x, y), in which two consecutive cells are adjacent (possibly diagonally). The result of combining the two fragments is illustrated in Fig. 5.13 for the case shown

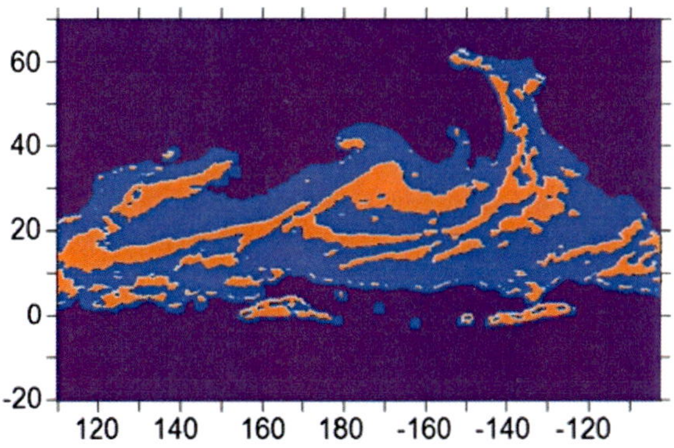

Fig. 5.13 The result of combining ridges

above in Fig. 5.12. The maximum order (the longest length of the connecting chain) $k_{\max}$ is set to 7. The search for the chain is carried out recursively. If there are several options for the connecting chain of equal length, the choice of solution is determined by the direction of the bypass of neighboring cells (clockwise).

In conclusion of this paragraph, it should be emphasized once again that the described morphological analysis procedure is performed twice: to search for longitudinally oriented and latitudinally oriented fragments of the AR, Fig. 5.14.

In the case of searching for latitudinally oriented fragments, instead of expressions (5.13) and (5.14), similar ones were used, with a variation of the first index of the cell instead of the second, which corresponds to the orthogonal direction of analysis compared to the one described above. The result of this analysis step is two sets of fragments: with longitudinally oriented ridges $C_i^{lon}(m_i, n_i)$ and with latitudinally oriented ones $C_j^{lat}(m_j, n_j)$.

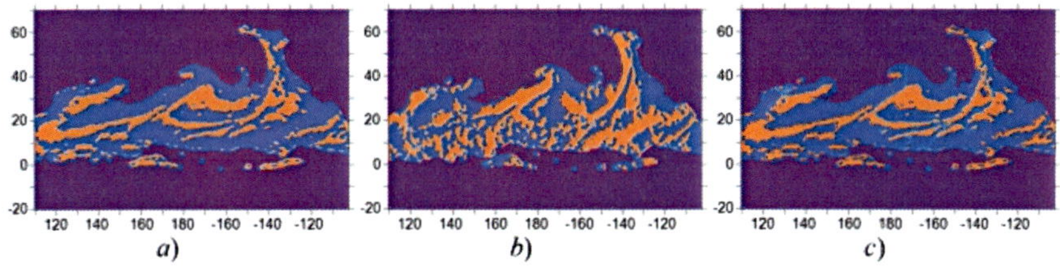

Fig. 5.14 **a** Longitudinally-oriented and **b** latitudinally-oriented ridges in the TPW field; **c** the result of their combination (the restored fragments of the latitudinally oriented ridges are shown in blue; the longitudinally oriented ridges in white)

5.2.4 Combining the Fragments

The step of **combining the fragments** is aimed at restoring the most complete images of the AR axes in the TPW field, the fragmentation of which is associated with a significant change in the direction of their flow (from longitudinal to latitudinal). As noted above, in automatic detection, preference is given to the search and localization of longitudinally oriented fragments. They form the basis of the reconstructed axes of the AR. However, at the same time, discontinuities arise in the images of the axes of the AR in areas where they have a predominantly latitudinal orientation. To eliminate these gaps, a search for latitudinally oriented fragments is carried out, which connect two longitudinally oriented fragments to each other. Algorithmically, this step is completely analogous to the step described above combining longitudinally oriented fragments, with the exception that in this case the connecting chain should be formed by cells belonging to one latitudinally oriented fragment. The result of the algorithm is illustrated by an example in Fig. 5.14c. The priority of the longitudinally oriented fragments is ensured by the fact that all of them at this stage of processing are stored as potential axes of the AR. In this case, the latitudinally oriented fragments are preserved only if two fragments of the longitudinal orientation are connected together. In the latter case, the longitudinally oriented fragments together with the corresponding latitudinally oriented ones form a new composite fragment.

5.2.5 Pruning Branches

The procedures for combining fragments lead to the formation of a series of false decisions related to the fact that some of the latitudinally oriented fragments that are not connected with the axes of the AR turn out to be connected to a pair of longitudinally oriented fragments and combine with them into a composite one. To eliminate a number of false decisions, an additional filtering operation, called branch pruning, has been implemented. For this purpose, a finite-difference analog of the Laplace operator is applied to the smoothed-up median filtering field of the TPW, which makes it possible to distinguish two-dimensional local maxima in the smoothed field. The value calculated in the cell (i, j) is denoted by $L(i, j)$.

For each analyzed composite (i.e., including latitudinally oriented ridges) of the AR fragment, the average $\overline{L}$ and variance D_L of values $L(i, j)$ for all cells of the fragment are calculated. Then, the cells included in the analyzed fragment as a connecting latitudinally oriented ridge are checked. If the value $L(m, n)$ in the cell (m, n) does not satisfy the requirement

$$L(m, n) \geq \overline{L} - 2\sqrt{D_L}, \tag{5.16}$$

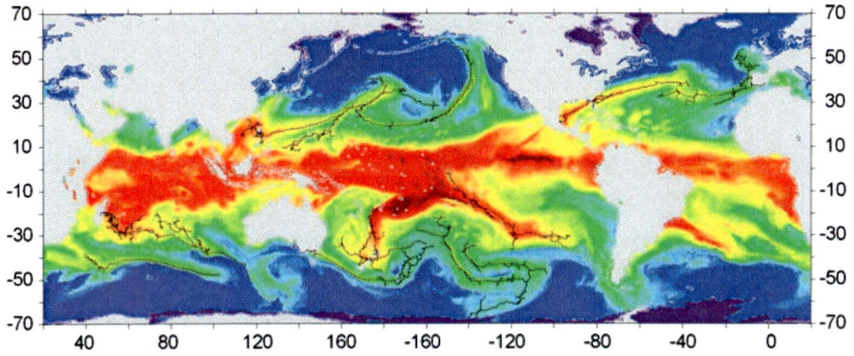

Fig. 5.15 The detected axes of atmospheric rivers (black lines) superimposed on the global field of the TPW of the atmosphere over the ocean (color scale—as in Fig. 5.1). The land mask is shown in gray. At the edges of the picture—geographical coordinates in degrees

then this cell is excluded from the analyzed composite fragment (as a result of which the fragment can again be divided into two parts). This procedure is repeated with all analyzed fragments.

At the final filtration step, a simple check of the linear dimensions and location of all remaining fragments is performed. If the length of the fragment in both dimensions does not exceed 80 cells (20 geographical degrees), the fragment is excluded from consideration as an axis of the AR. In addition, a fragment is excluded from consideration if it does not contain cells belonging to temperate latitudes, i.e. located entirely in the lower or entirely in high latitudes. This test is aimed at eliminating false decisions related to large-scale features of the TPW field in the intertropical convergence zone and in the frontal areas of extratropical cyclones, especially in the southern hemisphere.

As an example of processing, the final result of automatic detection of the AR axes in the TPW field over the Pacific, Indian and Atlantic oceans is shown in Fig. 5.15.

It can be seen that, in general, the implemented detection algorithm satisfactorily copes with the task of distinguishing the fine structure of the TPW field over the World Ocean. However, the question of whether all the detected filamentary structures are atmospheric rivers remains open. One of the difficulties of the question is that, in fact, no independent (from the phenomenological rules, one way or another used in the implemented algorithm) definition of the AR was proposed, which would allow verifying the results of the detection of the AR. Some additional important information can be obtained from the analysis of synchronous fields of other geophysical parameters (cloud liquid water content, precipitation rate) (Matrosov 2013), see also (Shields et al. 2018). An important aspect of the problem is how effectively the selected filamentary objects participate in the transfer of atmospheric latent heat. A systematic approach to solving these problems can also be implemented using satellite radiothermovision, as shown in the next section.

5.3 Retrieval of Atmospheric River Characteristics

The two previous sections of the chapter describe the specific problems that arise in the study of AR using satellite radiometric methods, and demonstrate the advantages of satellite radiothermovision, which can mainly overcome these problems and implement a systematic analysis of AR as a structural element of global atmospheric circulation.

As shown before, satellite radiothermovision provides the possibility of analyzing global TPW fields, which makes it possible to distinguish the fine structure of "filamentary formations" in them (Newell et al. 1992). However, this far from exhausts the possibilities of the proposed approach. Satellite radiothermovision provides the possibility of analyzing the processes under study in dynamics, calculating a number of integral characteristics (such as the latent heat flux power through arbitrarily set boundaries, including the meridional component of latent heat advection, see Chap. 3). Of great interest is the possibility of combining and joint analysis of several data series and/or fields of several geophysical parameters due to their mutual spatiotemporal combination.

Available satellite data are sufficient to study the role of AR in the global atmospheric circulation at climatically significant time intervals (of the order of 15 years, see Chaps. 6, 7). Carrying out this kind of large-scale research requires the realization of a number of stages: the implementation of streaming processing of remote data according to the algorithms described above; the selection and cataloging of all filamentary structures, potentially representing atmospheric rivers; applying to them further procedures for calculating integral characteristics (such as latent heat flux power); formation and analysis of statistics of these calculated values for the entire array of observational data.

In this section, as a brief conclusion, some of the indicated possibilities of such an analysis, provided by the approach of satellite radiothermovision, are illustrated. Full-scale studies according to the plan outlined above go beyond the scope of the book and represent one of the promising areas for further development in this area.

5.3.1 *Analysis of Latent Heat Fluxes*

As noted above, one of the reasons for the high interest in ARs is the result obtained in (Zhu and Newell 1998) on the basis of modeling, indicating that ARs can be a determining factor in the meridional transfer of atmospheric latent heat, at least in moderate latitudes. The frequency of formation and intensity of AR can significantly depend on the current state of the climate, and their variations can reflect global climate change. Therefore, studies that are able to clarify the role of AR in the general circulation of the atmosphere based on representative arrays of observational data are extremely relevant. Here, the appearance of works investigating the formation of

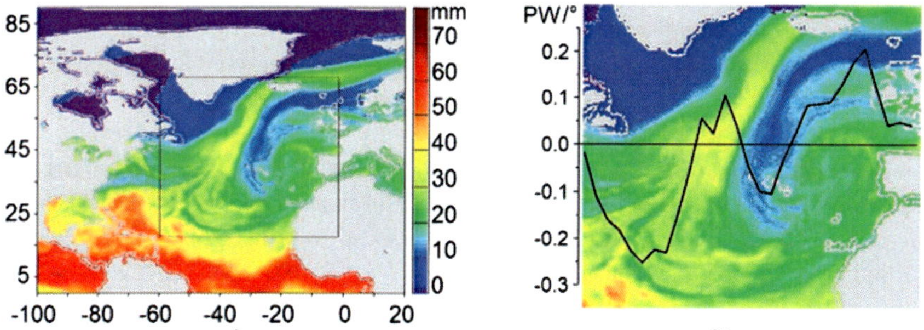

a) *b)*

Fig. 5.16 Advection of latent heat in an atmospheric river: **a** an atmospheric river in a TPW field above the North Atlantic on 10/21/2016, a color scale of TPW values on the right; **b** a graph of the power of the advective flux of latent heat across the 45° N boundary as a function of longitude superimposed on the corresponding fragment of the TPW field (marked with a black frame in Fig. 5.16a). See explanations in the text

AR or similar phenomena over land in temperate latitudes and their potential impact on the weather and climate of the Arctic (Komatsu et al. 2018) should be noted.

An algorithm for analyzing the filamentary structure of the TPW field and identifying potential axes of the AR was discussed in the previous section. The following example illustrates the capabilities of satellite radiothermovision for calculating the power of the meridional latent heat flux realized in the AR and comparing it with the "background" values of the flux power outside it, Fig. 5.16.

In the example shown in Fig. 5.16a, in the TPW field for October 21, 2016, the AR is clearly visible. The yellow–green strip, corresponding to TPW values of about 30–40 mm, crosses the North Atlantic, passes over Iceland, reaches the Arctic and continues between the Scandinavian peninsula and the Spitsbergen archipelago. To analyze its contribution to the total meridional transport of latent heat, a boundary is introduced that crosses the North Atlantic at 45° north latitude, and the latent heat flux through the boundary elements with a length of 1° longitude (in PW/°) is calculated. At a latitude of 45°, a value of 0.01 PW/° corresponds to approximately 127 MW/m.

Figure 5.16b illustrates the calculation results. For clarity, a graph of the flux power, as a function of longitude, is superimposed on the corresponding fragment of the TPW field (highlighted by a frame in Fig. 5.16a), and the horizontal axis (zero flux values) is aligned with a parallel of 45° N. The graph reveals the region of fluxes directed toward the pole with a maximum power of about 0.1 PW/° around the axis of the AR, the region of even more powerful flows in the north direction, formed by the remnants of another AR, involved in strong cyclonic rotation west of the Iberian Peninsula. These two areas are separated by the invasion of cold air from the Arctic, where the flux changes direction (which corresponds to a change in the power sign in the graph below to negative). Another cold intrusion is also observed, coming from the Labrador Sea and expressed in even more powerful latent heat flux in a southerly direction due to the relatively high TPW values at the 45° boundary.

This example demonstrates the high promise of using satellite radiothermovision to study atmospheric rivers, including as a climate-forming factor. The uniqueness of the approach is that the calculation results are based entirely on satellite observation data and can be considered independently of the calculation results from circulation models. It should be noted that, as in the study of the evolution of tropical cyclones, a qualitative improvement in the analysis should be expected with the additional use of AMSR-2 data of higher spatial resolution using multisensory satellite radiothermovision algorithms (Chap. 3).

5.3.2 Analysis of AR Images in Dynamics

The automatic detection and localization of ARs in the TPW field, as a special case of any detection and localization task, is almost inevitably associated with the adoption of two types of erroneous decisions: the false classification of some features of the TPW field as an AR and the omission of the AR image in the TPW field. One of the sources of errors is the tuning parameters of the algorithms, the values of which are established empirically based on preliminary analysis of a limited set of data, which may turn out to be insufficiently representative. In addition, (including due to the insufficient rigor of the very concept of AR and corresponding detection criteria), there may be cases when two fragments of one AR will be recognized as two ARs, or, conversely, the images of two ARs will be combined and recognized as one AR of a more complex shape.

And here, some additional information in support of a particular solution can be given by analyzing the TPW field in dynamics, revealing the chronological sequence of phases of the evolution of AR. This is illustrated by the example in Fig. 5.17.

Figure 5.17 shows the fragments of the TPW field over the North Pacific Ocean in dynamics: the fragments in Fig. 5.17a–e relate to January 1, 2016 and were constructed in 3 h increments (for local time moments around 6:00, 9:00, 12: 00, 15:00, 18:00, 21:00). Further, to illustrate a longer-term evolution, fragments of the TPW field for January 3 and 10, 2016 are shown. It can be seen that the ridges corresponding to the axes of the potential ARs are mainly correctly identified in the TPW field as a result of their automatic detection. In this case, however, the filamentary structure of the TPW field changes morphologically significantly. Indeed, in Fig. 5.17a, three potential ARs are visible, which can be arbitrarily called "western", "central" and "eastern". In Fig. 5.17b, the axes of the western and central ARs merge into a single object. In Fig. 5.17c, the axis of the eastern AR (filtered out by the length criterion) is not detected. At the same time, the combined "western" and "central" ARs receive two different "sources". In Fig. 5.17d, a situation similar to that obtained in Fig. 5.17b is restored, but the formation of a new AR along the western coast of the Pacific Ocean is additionally discovered. In Fig. 5.17d, the axis of the new AR merges with the initial fragment of the axis of the previously combined "western" and "central" AR, which, in turn, is separated from their continuation over the central region of the Pacific Ocean. This continuation also combines the eastern AR.

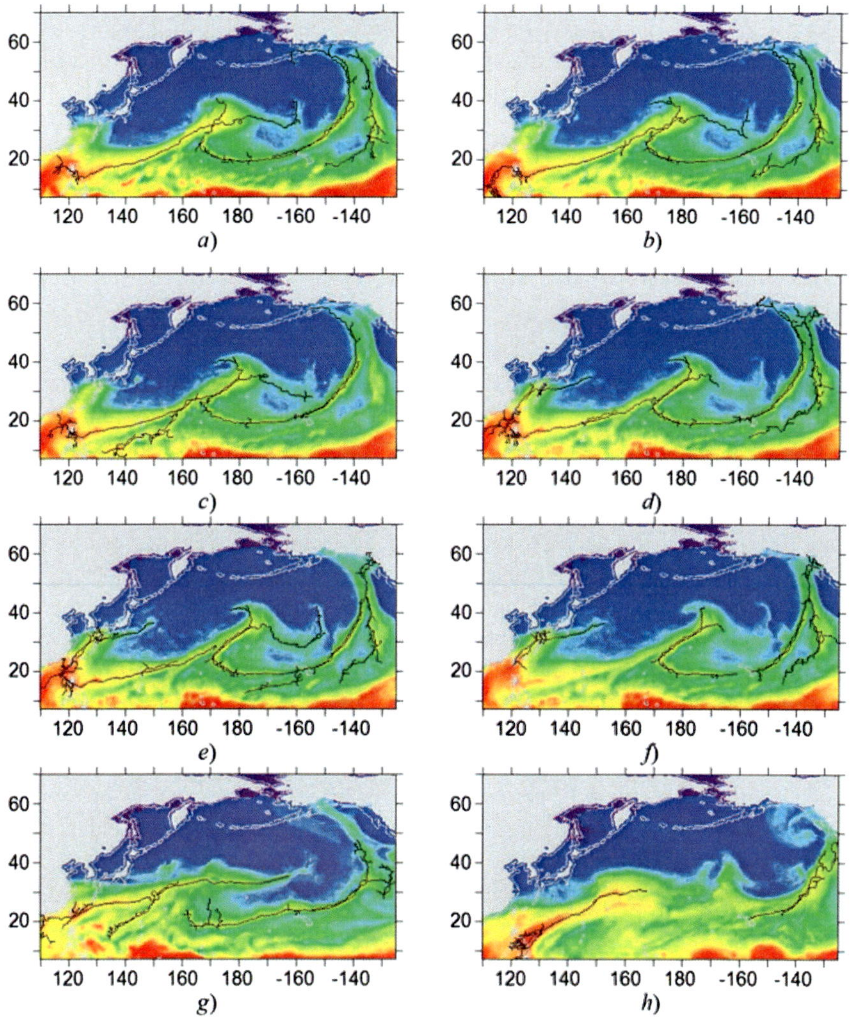

Fig. 5.17 Phases of evolution of atmospheric rivers in the TPW field over the North Atlantic: **a–f** 01/01/2016; **g** 01/03/2016; **h** 01/10/2016. The color scale of TPW values as in Fig. 5.1

In Fig. 5.17e, the images of the combined "western" and "central" ARs decay to a large extent. The new AR continues to exist as an independent entity. Fragments related to January 3 and 10 illustrate the further evolution of the TPW field in the considered area. It is clear that the combined analysis of Figs. 5.17a–e is able to more accurately establish the exact number of ARs and the configuration of their axes, taking into account the short-term history of the evolution of the TPW field in this fragment. In this case, however, a sufficiently high detailing in time is required. So, the detected AR images in Fig. 5.17g (January 3) are difficult to reliably associate with any AR images in Fig. 5.17e (January 1). On January 10, all detected axes belong to ARs that did not exist on January 1 and 3. Thus, the capabilities of satellite radiothermovision in terms of studying short-term atmospheric dynamics contribute to increasing the accuracy of automatic analysis of the fine structure of the TPW field.

5.3.3 *Joint Analysis in the Fields of Several Geophysical Parameters of the Atmosphere*

As noted above, a joint analysis of the fields of several geophysical parameters opens up additional possibilities. Thus, ARs are characterized not only by high gradients of the TPW field across the axes, but also by relatively high cloud cover and intense precipitation (Matrosov 2013). In this regard, it is advisable to consider the possibility of detecting ARs in the combined (in space and time) fields of the TPW, the cloud liquid water content, the precipitation rate, etc. By analogy with (Liu 1988; Liu and Tang 2005) it is also interesting to take into account in the analysis of the surface wind speed, which to some extent characterize advection in the lower troposphere. The combination of these fields and their calculation with the necessary time detail can be carried out using satellite radiothermovision algorithms. As an example, Fig. 5.18 shows the fields of the TPW Q, the cloud liquid water content L, and the surface wind speed W for December 25, 2016.

A global field of magnitude $P = Q \cdot L \cdot W$ is also given in relative units, normalized to a value range of 0–250. The combination of Q, L, and W fields in the form of a product, offers the option of combining fields of different physical nature and/or having significantly different ranges of variability, making it possible to balance the contributions of spatial contrasts of all components.

Indeed, for $P \neq 0$:

$$\frac{1}{P}\frac{\partial P}{\partial x} = \left(\frac{\partial Q}{\partial x}LW + \frac{\partial L}{\partial x}QW + \frac{\partial W}{\partial x}QL\right)/QLW = \frac{1}{Q}\frac{\partial Q}{\partial x} + \frac{1}{L}\frac{\partial L}{\partial x} + \frac{1}{W}\frac{\partial W}{\partial x}, \tag{5.17}$$

where x is the spatial coordinate. As can be seen in Fig. 5.18d, the "filamentary" structure of global circulation, including potential ARs, is clearly manifested in the P field.

5.3.4 *Joint Analysis Over the Ocean and Land*

Substantial progress in understanding the nature of AR can be ensured by the study of their evolution over land, where, due to significantly lower, on average, TPW values, the main mechanism of their formation and movement is the advection of oceanic

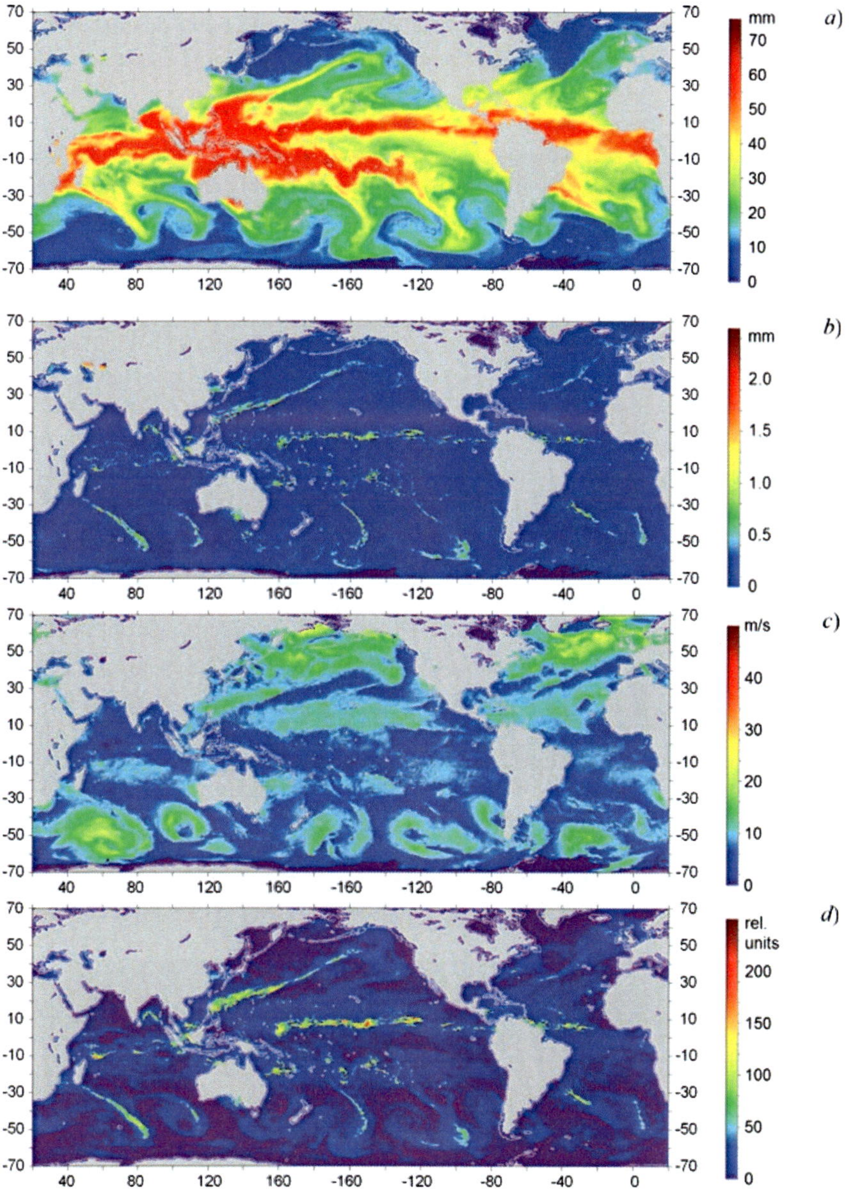

Fig. 5.18 Global fields of total precipitable water Q, mm (**a**); cloud liquid water content L, mm (**b**); surface wind speed W, m/s (**c**) and their product P, rel. units (**d**) based on satellite radiometry remote data for December 25, 2016. At the edges of the images—geographical coordinates in degrees. Color scales of field values are to the right of the corresponding images

air masses. Although the restoration of the TPW field over land encounters serious difficulties, lately some progress has been achieved in this area (Du et al. 2017) associated with the use of polarimetric multichannel measurements. To date, such measurements have been carried out only by Japanese devices of the AMSR series (AMSR-E, AMSR-2) and Russian devices of the MTVZA series (MTVZA-GYa). From the point of view of obtaining long-term stable, reliably calibrated data series, preference should be given to the instrument of the AMSR series. Unfortunately, unlike the SSM/I series of devices, the AMSR-E (2002–2011) and ASMR-2 (2012–2018) devices operated in satellite orbits in a single instance each. This does not allow the full application of satellite radiothermovision algorithms to their data. However, TPW fields over the ocean according to SSM/I, SSMIS, WindSat, etc., using spatiotemporal interpolation carried out in satellite radiothermovision, can be combined in time with the corresponding TPW fields over land according to AMSR-E and AMSR-2, which ensures the construction of a joint TPW field with the most complete global coverage. An example of such a combined TPW field is shown in Fig. 5.19 and relates to the measurement data for June 28, 2013.

As you can see, the TPW field above the ocean does not contain lacunae. The TPW field over land is determined only at the cells in which measurements were made with the AMSR-2 instrument (within its swaths). At the same time, satellite radiothermovision algorithms allow efficient combination of data over time. Indeed, it is clearly seen that the images of the ARs and other features of the TPW field do not experience sharp discontinuities when crossing the land-ocean boundaries. Figure 5.19 shows several ARs that cross significant areas of the atmosphere above land and pass from one ocean to another (from the Indian to the Pacific and from the Pacific to the Atlantic). In this context, we should again pay attention to the direction of research outlined in (Komatsu et al. 2018). It should be expected that the study of such a most complete picture in dynamics will open up new prospects in the investigation of the nature of AR. At the same time, the simultaneous operation

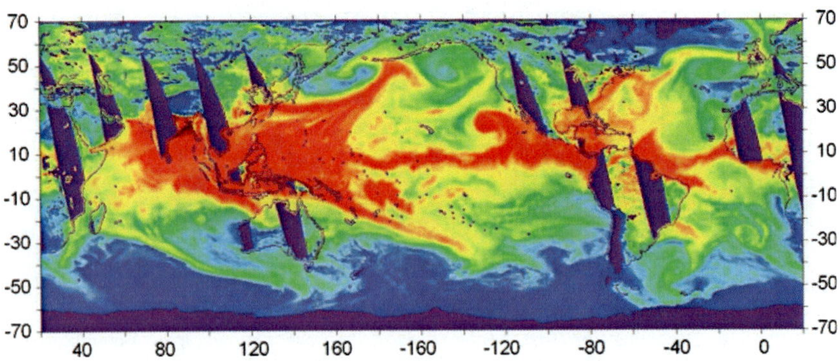

Fig. 5.19 Global TPW field over the ocean and land according to the SSM/I, SSMIS, WindSat and ASMR-2 data for June 28, 2013, combined using satellite radiothermovision algorithms. The color scale of TPW values is as in Fig. 5.1. At the edges of the image, the geographical coordinates in degrees. The boundaries of the continents and areas of data gaps are indicated by a purple mask

in orbit of several devices of the type AMSR-2 or MTVZA-GY will provide (in the presence of long-term, stable, continuous, well-calibrated measurements) the calculation of the total TPW field both over the ocean and over land, without gaps and with high detail (no worse than 3 h) in time.

5.4 Concluding Remarks to this Chapter

1. The results presented in this chapter relate to applications of the satellite radiothermovision approach to studies of the formation and evolution of synoptic atmospheric processes using atmospheric rivers as a very important example.
2. The proposed methodology for processing satellite radiometry remote sensing data, based on a set of implemented satellite radiothermovision algorithms, allowed to overcome the main difficulties in analyzing the filamentary structure of global atmospheric circulation and atmospheric river research: fragmentation of objects of research due to their significant zonal extent and divergence of satellite swaths; spatiotemporal combination of fields of various geophysical parameters for their joint analysis; the need to obtain quantitative information on the flows of latent heat generated by ARs.
3. An algorithm for automatic detection of ARs and AR-like atmospheric processes is proposed and implemented in software. The computation is performed and the power of latent heat fluxes realized in the AR and the "background fluxes" outside the AR are compared. The possibilities of performing a joint analysis of synchronized fields of several geophysical parameters of the ocean—atmosphere system are demonstrated to increase the reliability of automated detection of ARs and study their evolution.

References

Bao JW, Michelson SA, Neiman PJ, Ralph FM, Wilczak JM (2006) Interpretation of enhanced integrated water vapor bands associated with extratropical cyclones: their formation and connection to tropical moisture. Mon Weather Rev 134(4):1063–1080

Dettinger MD, Ralph FM, Das T, Neiman PG, Cayan DR (2011) Atmospheric rivers, floods and water resources of California. Water 3(2):445–478

Du J, Kimball JS, Jones LA, Kim Y, Glassy J, Watts JD (2017) A global satellite environmental data record derived from AMSR-E and AMSR2 microwave earth observations. Earth Syst Sci Data 9(2):791–808

Ermakov DM (2017) Investigation of the Features of long-term global atmospheric circulation via satellite radiothermovision. In: Proceedings of 2017 progress in electromagnetics research symposium—Spring (PIERS), pp 413–418. https://doi.org/10.1109/piers.2017.8261775

Ermakov DM (2019) Development of the climatological database of atmospheric rivers. In: Proceedings of the V international conference information technologies in earth sciences and applications for geology, mining and economy. ITES&MP—2019, Moscow (Russia), 14–18 October 2019. VNIIgeosystem, Moscow, p 61

Ermakov DM, Chernushich AP (2018) Development of automatic algorithms for detecting atmospheric rivers, In: Information technologies in remote sensing of the earth—RORSE 2018, pp 68–75. https://doi.org/10.21046/rorse2018.68. (in Russian)

Ermakov DM, Sharkov EA, Chernushich AP (2015) Satellite radiothermovision of atmospheric mesoscale processes: case study of tropical cyclones. In: The international archives of the photogrammetry, remote sensing and spatial information sciences—ISPRS archives, vol XL(7/W3), pp 179–186

Ermakov DM, Sharkov EA, Chernushich AP (2016) A multisensory algorithm of satellite radiothermovision. Izvestiya Atmos Oceanic Phys 52(9):1172–1180

Ermakov DM, Sharkov EA, Chernushich AP (2017) Satellite radiothermovision on synoptic and climatically significant scales. Izvestiya Atmos Oceanic Phys 53(9):973–978

Ermakov DM, Sharkov EA, Chernushich AP (2019) Evaluation of tropospheric latent heat advective fluxes over the ocean by the animated analysis of satellite radiothermal remote data. Izvestiya Atmos Oceanic Phys 55(9):1125–1132. https://doi.org/10.1134/S0001433819090160

Gimeno L, Nieto R, Vazquez M, Lavers DA (2014) Atmospheric rivers: a mini-review. Frontiers Earth Sci 2. https://doi.org/10.3389/feart.2014.00002

Knippertz P, Wernli H (2010) A lagrangian climatology of tropical moisture exports to the Northern Hemispheric extratropics. J Clim 23(4):987–1003

Komatsu KK, Alexeev VA, Repina IA, Tachibana Y (2018) Poleward upgliding Siberian atmospheric rivers over sea ice heat up Arctic upper air. Sci Rep 8:2872. https://doi.org/10.1038/s41598-018-21159-6

Liu WT (1988) Moisture and latent heat flux variabilities in the Tropical Pacific derived from satellite data. J Geophys Res 93(C6):6749–6760

Liu WT, Tang W (2005) Estimating moisture transport over oceans using space-based observations. J Geophys Res 110(D10):D10101. https://doi.org/10.1029/2004JD005300

Matrosov SY (2013) Characteristics of landfalling atmospheric rivers inferred from satellite observations over the Eastern North Pacific ocean. Mon Weather Rev 141(11):3757–3768

Nerushev AF, Kramchaninova EK (2011) Method for determining atmospheric motion characteristics using measurements on geostationary meteorological satellites. Izvestiya Atmos Oceanic Phys 47(9):1104–1113

Nerushev AF, Visheratin KN, Ivangorodsky RV (2018) Spatiotemporal variability of high-altitude jet streams from satellite measurements. Izvestiya Atmos Oceanic Phys 54(9):1076–1088

Nerushev AF, Visheratin KN, Ivangorodsky RV (2019) Dynamics of high-altitude jet streams from satellite measurements and their relationship with climatic parameters and large-scale atmospheric phenomena. Izvestiya Atmos Oceanic Phys 55(9):1198–1209

Newell RE, Newell NE, Zhu Y, Scott C (1992) Tropospheric rivers?—a pilot study. Geophys Res Lett 19(24):2401–2404

Palmén E, Newton CW (1969) Atmospheric circulation systems: their structural and physical interpretation, vol XVIII. Academic Press. New York, 1969, p 606

Ralph FM, Neiman PJ, Wick GA (2004) Satellite and CALJET aircraft observations of atmospheric rivers over the eastern North Pacific Ocean during the winter of 1997/98. Mon Weather Rev 132(7):1721–1745

Shields CA et al (2018) Atmospheric river tracking method intercomparison project (ARTMIP): project goals and experimental design. Geosci Model Dev 11(6):2455–2474. https://doi.org/10.5194/gmd-11-2455-2018

Wick GA, Neiman PJ, Ralph FM (2013) Description and validation of an automated objective technique for identification and characterization of the integrated water vapor signature of atmospheric rivers. IEEE Trans Geosci Remote Sens 51(4):2166–2176

Zhu Y, Newell RE (1994) Atmospheric rivers and bombs. Geophys Res Lett 21(18):1999–2002

Zhu Y, Newell RE (1998) A proposed algorithm for moisture fluxes from atmospheric rivers. Mon Weather Rev 126(3):725–735

Chapter 6
Satellite Radiothermovision of Global Atmospheric Circulation

The study of global atmospheric circulation is, of course, closely connected with the task of studying the Earth's climate and its changes, which, in turn, is of not only fundamental scientific, but also universal interest.

The most important tool for studying climate today is mathematical modeling. It not only provides a qualitative picture of the evolution of climate over epochs spanning millions of years, but also makes it possible (many times!) to set up numerical experiments on climate models, evaluating the influence of various disturbing influences. However, it should be borne in mind that the forecast of climate change in numerical models is a family of "scenarios"—possible trajectories of a point in the phase space of climatic variables that wander around a certain attractor and sometimes pass from one attractor to another, e.g. (Barry and Chorley 2003; Henderson-Sellers and McGuffie 2012). The choice of a particular scenario by the climate system depends, among others, on a large set of random small-scale factors, the deterministic accounting of which is impossible in principle. To clarify the current climate evolution scenario, real-world observations are needed, reflecting not only its "instantaneous" state, but also short-term dynamics. It should be expected that the need for such data will only increase with the development and refinement of climate models.

In this general context, this review is devoted to what qualitative and quantitative information about the climate system can be extracted based on satellite radiothermal monitoring of the Earth using the satellite radiothermovision approach. Of greatest interest is the joint use of fields of scalar geophysical quantities and the corresponding vector fields of advection for calculating the integral physical characteristics of processes, in particular, horizontal flows of latent heat through predetermined boundaries. The chapter demonstrates the possibilities and some results of applying this approach in studies of the current state and evolution of the atmospheric circulation of latent heat on planetary scales and climatically significant time intervals. These issues are discussed in more detail in publications (Ermakov 2017, 2018; Ermakov et al. 2017a).

D. M. Ermakov, *Satellite Radiothermovision of Atmospheric Processes*, Springer Praxis Books,
https://doi.org/10.1007/978-3-030-57085-9_6

6.1 Characteristics of the Problem in the Light of Satellite Radiothermovision

Atmospheric transport of apparent and latent heat is one of the main factors in the formation of the weather and climate of the Earth. So, on a climatic scale, the contribution of meridional heat transfer by the atmosphere is comparable to the contribution of the ocean in the tropics and exceeds it in the middle and high latitudes (Trenberth and Caron 2001; Wunch 2005). Moreover, if the apparent heat flux is everywhere directed towards the poles, then the latent heat flux in a first approximation reproduces the global circulation pattern, forming an intratropical convergence zone with a meridional transport component directed to the equator. At the same time, in the structure of latent heat fluxes there is no marked boundary between polar cells and Ferrel cells, i.e. in temperate and high latitudes, the average latent heat flux is directed to the poles (Palmén and Newton 1969). It should be noted that model estimates of the meridional heat transfer by the climate system contain uncertainties associated, in particular, with insufficient knowledge of the general radiation balance of the Earth (Trenberth and Caron 2001; Wunch 2005). This uncertainty must be borne in mind when dividing the total transfer into components, the contribution of which can be comparable with the magnitude of the errors.

At synoptic scales, rapidly developing atmospheric processes can introduce significant disturbances into the general circulation structure. One of the important factors of meridional transfer of latent heat is currently considered the so-called atmospheric rivers (Newell et al. 1992; Zhu and Newell 1994, 1998), see Chap. 5. Forming over the oceans in a humid tropical atmosphere, they cross the boundaries of Headley cells and carry out rapid transport of water vapor to middle and high latitudes in the northeast (in the northern hemisphere) and southeast (in the southern hemisphere) directions. The connection of atmospheric rivers with extreme weather events (torrential rains, floods, mudflows) on the western coast of North America is proved. Similarly, atmospheric rivers over the Atlantic can have a significant impact on the weather of Western and Northern Europe and the Arctic.

An impressive example of the existence of "telecommunications" in the Earth's atmosphere is the exchange of latent heat between the southern and northern hemispheres. According to observational data and model calculations, water vapor condensation (precipitation) prevails over evaporation in the northern hemisphere, while the opposite situation (prevalence of evaporation) takes place in the southern hemisphere. So, according to the estimates given in (Palmén and Newton 1969) partly based on (Budyko 1956, 1963; Sellers 1965), in order to maintain balance, it is necessary that the equator annually crosses northward 1.647×10^{16} kg of water vapor, which is equivalent to a latent heat flux of 1.2 PW. This, of course, is one of the consequences of the general asymmetry of the Earth's climate system (Feulner et al. 2013; Kang and Seager 2015).

Another important process that clearly demonstrates the asymmetry of atmospheric circulation is global tropical cyclogenesis (Emanuel 2003; Golitsyn 2008; Sharkov 2000, 2012). In general, the trajectories of tropical cyclones are determined by the general nature of the atmospheric circulation over the corresponding basin of the World Ocean and the current state of the geophysical fields of the ocean—atmosphere system. Originating and receiving energy in tropical regions, cyclones, as a rule, drift westward with a component to high latitudes, and, reaching moderate latitudes, move eastward, also with a component to high latitudes, dissipating (filling) releasing the stored energy in environment. Of course, global tropical cyclogenesis is in a complex relationship with the climatic regime of the Earth. Moreover, its activity in the northern hemisphere is higher. A certain analogy can be noted between atmospheric rivers causing extreme weather events on the western coasts of the continents and tropical cyclogenesis, which creates constant threats of atmospheric catastrophes on the eastern coasts, and the short-term impact on the local ecosystem and infrastructure from typhoons and hurricanes is often accompanied by much more serious local consequences.

This chapter demonstrates that the calculation of latent heat fluxes from satellite thermal data from global observations can be carried out with detail on the order of a fraction of a degree and a time step of units of hours. The results of these calculations can be used as an additional source of information on the energy balance in the so-called "subgrid processes" of modern climate models, the spatial resolution of which is now of order of a degree or half-degree (Taylor et al. 2012; Eyring et al. 2015; Volodin et al. 2017). The described approach can also be considered as independent of numerical climate modeling, an additional tool for monitoring and studying climate variations.

A basic approach to reconstructing the dynamics of geophysical fields is described in Chap. 3, based on publications (Ermakov et al. 2011, 2016b, 2019a, b) and a number of others. A generalized algorithm for estimating latent heat fluxes used to calculate meridional and zonal fluxes is described in more detail in (Ermakov et al. 2017a). The following sections describe the data used and the methodology for their processing and analysis, and discuss the results.

It should be noted that there are a number of alternative "synthetic" approaches in which the computational scheme is constructed (or tuned) with the use of additional information (radiosonde measurements, reanalysis results, etc.), but then uses only remote sensing data as input. So, in (Liu and Tang 2005), a neural network approach was proposed to restore the effective velocity and direction of advection of water vapor by scatterometric measurements of the drive wind speed and a number of additional parameters (time, geographical coordinates of the place, etc.). As a result, in spite of the known problematic aspects, it was possible to obtain realistic estimates of the average seasonal meridional and zonal fluxes of latent heat in the latitudinal belt from 40° S up to 40° N in the observation interval from August 1999 to August 2003 (ibid.). A brief overview of "synthetic" type approaches can be found in (Robertson et al. 2014).

6.2 Used Data and Analysis Technique

Total precipitable water (TPW) fields of the atmosphere, freely distributed by Remote Sensing Systems, USA, were used as input information in the calculations (Wentz et al. 2012, 2013, 2014a, b), reconstructed from the data of SSM/I (DMSP F13, F14), SSMIS (DMSP F16, F17, F18), WindSat (Coriolis), as well as, in some cases, AMSR-E (Aqua), AMSR2 (GCOM-W1) and SSM/I (DMSP F15) measurements, see (Ermakov et al. 2016a). All calculations were performed on a regular grid of global coverage in increments of 0.25°. Interpolation was carried out in a continuous observation interval from 01/01/2003 to 10/01/2017 with a time step of 3 h, which ensured the calculation of advection fields in increments of 6 h (four times a day).

In general, the calculation results covered an almost complete 15-year time interval, forming about 22,000 pairs of scalar fields of integral moisture content with dimensions of 1440 × 720 elements and the corresponding vector fields of advection. Figure 6.1 illustrates examples of such pairs relating to the extreme dates of the entire range, 01/01/2003 and 10/01/2017, respectively. The color tone encodes the value of the TPW (the scale in mm is given below). Vectors indicate the direction and speed of advection. Due to projective distortions in the images, it is impossible

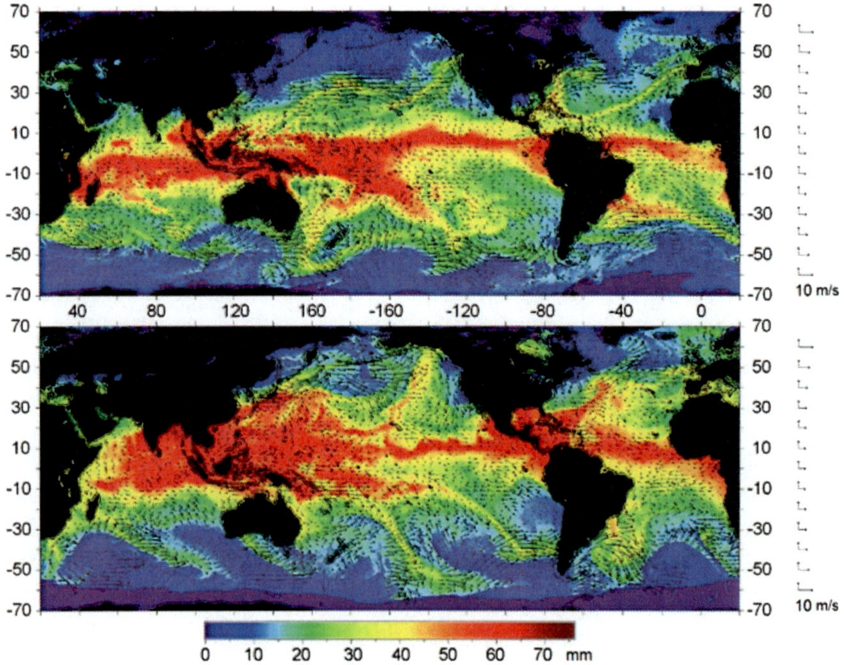

Fig. 6.1 Combined fields of total precipitable water (color scale in mm at the bottom) and advection rates (vectors, calibration standards on the right): at the top—calculation according to data on 01/01/2003; below—on 10/01/2017. At the edges of the images, the latitudes and longitudes in degrees

to observe a single display scale of the zonal component, therefore, the reference unit vectors corresponding to a speed of 10 m/s at different latitudes are shown on the right.

As a result of this preparatory phase of the study, in fact, a typologically new product is formed, which is a dynamic description of the Earth's atmosphere on a global scale over a long period of time (see Chap. 7). In addition to the problems considered below, it can be used, for example, in studies of the evolution of extratropical cyclones, dynamic and energy characteristics of atmospheric rivers (Wick et al. 2013; Ermakov 2017), changes in the zonal circulation index (Haltiner and Martin 1957), and many others. At the same time, it should be noted that the properties of the objects of study should be carefully correlated with the degree of detail of the obtained dynamic description. So, it was previously established (see Chap. 4) that an adequate description of the evolution of tropical cyclones requires calculations on a more detailed coordinate grid (0.125 or 0.2°) with a time step of 1.5 h or less (Ermakov et al. 2015, 2017b). In this case, the calculation procedure is significantly complicated, and more stringent requirements are imposed on the input data, which cardinally complicates the stream processing of large amounts of information.

The chapter demonstrates the application of the satellite radiothermovision approach for the analysis of meridional and zonal atmospheric circulation at global and regional scales and climatically significant time intervals. For this purpose, various ways of summarizing the results of preliminary processing are considered. In particular, seasonal and annual averaging of the fields of TPW and advection were applied. The general picture of atmospheric circulation obtained allows direct comparison with theoretical concepts and data of independent measurements. However, it gives a mostly qualitative description. For a more detailed quantitative description, heat fluxes through a specially organized grid of boundaries were calculated, which made it possible to focus the analysis on individual aspects of atmospheric circulation.

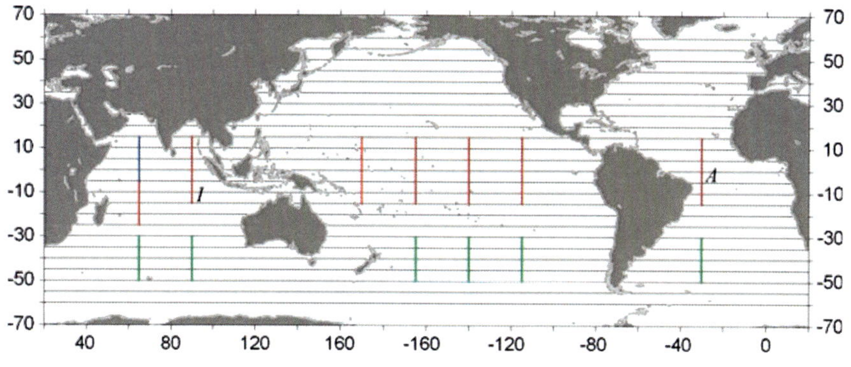

Fig. 6.2 Border families for calculating latent heat fluxes: black—along parallels with a step of 5°; colored—along the meridians. The meridional boundaries marked with the letters "I" and "A" are used in the calculations illustrated in Fig. 6.6

The boundary grid used in the analysis is illustrated in Fig. 6.2. To analyze the meridional circulation, were used the boundaries laid along the parallels from the equator to 60° latitude in both hemispheres with a step of 5° and shown in thin lines in Fig. 6.2. Similarly, for the analysis of zonal circulation, families of boundaries passing in the meridional direction and shown by thick colored lines have been introduced.

Due to the ability to perform calculations both over the entire length of the border and for its individual fragments, time series of latent heat fluxes were obtained both over the entire World Ocean and over individual basins. The possibility of an arbitrary choice of the position of the boundaries provided the "focus" of the analysis in space. Data samples and special time series processing techniques (such as wavelet analysis) provided focusing on time.

6.3 Analysis of the Retrieved Global Circulation Characteristics

To obtain the most general idea of the calculated data array, the average TPW field for the entire time of observations and the corresponding average advection field were constructed. The result is shown in Fig. 6.3. As before, color encodes TPW values; vectors display the advection field. It is seen that the zonal component dominates in the average circulation pattern. In accordance with well-known data (Palmén and Newton 1969), there is a stable eastern transport at low latitudes and an opposite western transport at higher latitudes up to 60–70°. The bands of near-zero velocity values at about 25° latitude in both hemispheres are in good agreement with the

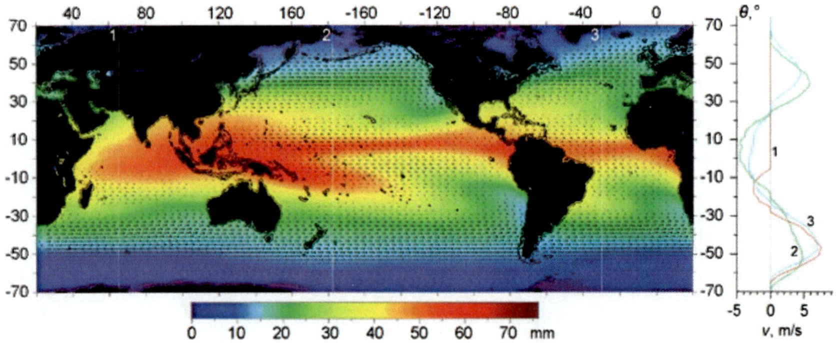

Fig. 6.3 Average field of integral atmospheric moisture content over the entire observation period, excluding incomplete 2017 (color scale in mm below) and the corresponding average field of advection velocities (vectors). On the right are the zonal components of the advection velocity v, m/s as a function of latitude θ along the meridional Section 1: 65° E; 2: 178° W; 3: 30° W in the range of 70° S up to 70° N

boundaries of the Headley cells. The greatest asymmetry of zonal currents is expected to be observed over the Indian Ocean.

Figure 6.3 shows three meridional sections laid over the waters of the Indian, Pacific and Atlantic oceans and marked with "1", "2" and "3". The change in the zonal component of the advection velocity v, m/s along these sections (as a function of latitude θ) is illustrated by curves 1–3 shown in the graph on the right. The relatively low maximum on curve 2 is due to the inhibitory effect of New Zealand and Australia. The results obtained are in good numerical agreement with the known data on the average zonal circulation. So, according to Fig. 1.2 from (Palmén and Newton 1969), the eastern transport in the lower troposphere of tropical latitudes occurs at a speed of about 5 m/s. The western transport at 40° latitude in both hemispheres is characterized by a range of average speeds from 2.5 m/s at the surface to 10 m/s at an altitude of about 3 km.

The meridional component of motion is relatively weakly expressed in the average circulation pattern, which in itself is in good agreement with the data shown in Figs. 1.2–1.4 from (Palmén and Newton 1969), from which it is seen that meridional advection is 2–4 times slower on average than zonal one. An additional decrease in the meridional component is associated with averaging of seasonal atmospheric fluctuations. The meridional transport is most pronounced in the middle latitudes near the western borders of the continents of both hemispheres. Its reason, obviously, is the interaction of the prevailing flows of Western transport with the continents. It can be noted that this peculiarity of circulation also manifests itself in the average TPW field in the form of "tongues" of relatively dry air, elongated from mid-latitudes to the equator over the eastern oceans. Another significant feature of middle meridional advection is the southern transport over the northeast Pacific Ocean and, especially, over the North Atlantic. This component of circulation is one of the main factors shaping the climate of the Arctic.

To study the effect of seasonal fluctuations on the average circulation pattern, it is reasonable to divide the entire data set into groups related to the warm and cold seasons of both hemispheres. Further, for brevity, "summer" will be called the three-month interval centered relative to June 22, and "winter"—the same interval from the middle of December 22, i.e. "boreal" summer and winter correspondingly. It is clear that "winter" conditionally corresponds to the warm season in the southern hemisphere, and "summer"—to the cold, and vice versa for the northern hemisphere.

The picture of average seasonal atmospheric circulation is illustrated in Fig. 6.4 using the example of summer 2003 (a) and winter 2003/2004 (b). Each season corresponds to a pair of images. The upper ones are constructed similarly to Fig. 6.3 and combine the middle fields of the TPW (color scale on the right) and advection (vectors). In the lower images, for the best visualization of the advection field, the absolute value of the speed is encoded in color (on the right is the range scale).

The geometric structure of the circulation is better seen in the upper images (against the background of the TPW fields). When divided into seasons, the picture of the average meridional circulation is much richer. In particular, large-scale anticyclonic rotations of air masses over the Pacific and Atlantic oceans (PO and AO), more pronounced during the warm season in each of the hemispheres, are observed (cf.

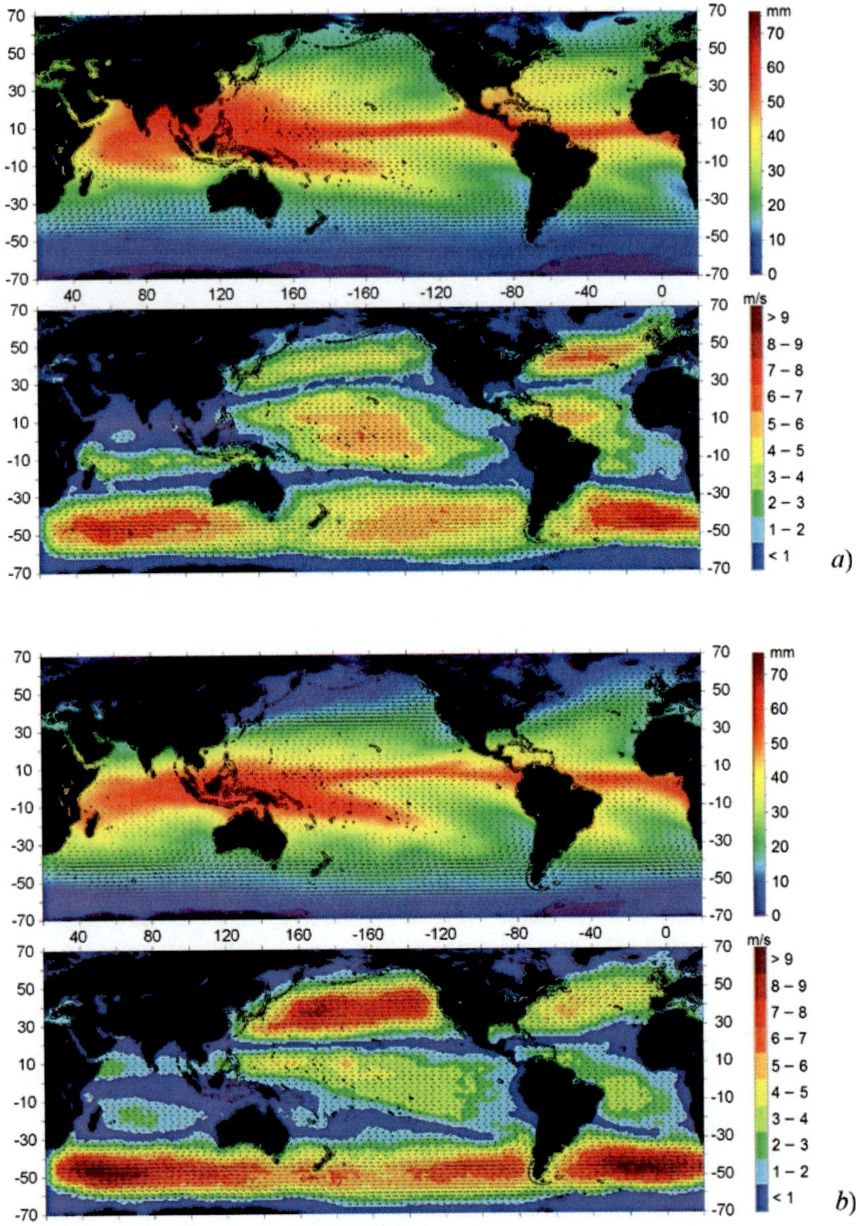

Fig. 6.4 Seasonal average fields of integral moisture content and advection rates: **a** summer 2003; **b** winter 2003/2004 (see explanations in the text)

Figs. 3.1–3.4 from (Palmén and Newton 1969)). The manifested pattern of circulation over the Indian Ocean (IO) also attracts attention.

Advection velocity maps (lower images in pairs) provide additional useful information. Thus, the boundaries of the Handley cells (HC) are clearly visible on them, especially above the PO and AO (regions of low wind speeds, colored in blue). In summer, the northern boundary of the HC noticeably deviates from the strictly zonal direction and is located between 20 and 30°, penetrating most deeply into the north above the central regions of PO and AO. The described circulation features are in good agreement with the above estimates (Liu and Tang 2005, Fig. 6.5a, JJA). Over IO, the region of weak advection is eroded to almost 5° south latitude. At the same time, the southern border of the HC almost everywhere passes around 25° south latitude with slight deviations from the parallel.

In winter, the situation changes to the opposite. The northern border of the HC lies along 20° north latitude over all oceans. The southern border of the HC is stretched by bands between 20 and 30°, with a stable tendency to shift to the south in the eastern parts. Here again, there is a clear agreement with the estimates (ibid., Figure 6.5a, DJF). A complex flow structure is formed above the IO, in which the region of weak advection forms an anticyclonic ring reaching the equator.

Further refinement of the atmospheric circulation characteristics is possible by calculating latent heat fluxes through the previously introduced boundary families (Fig. 6.2). Figure 6.5 shows an example of an analysis of latent heat fluxes over the oceans across the borders above the equator and 25° north latitude. In the graph of

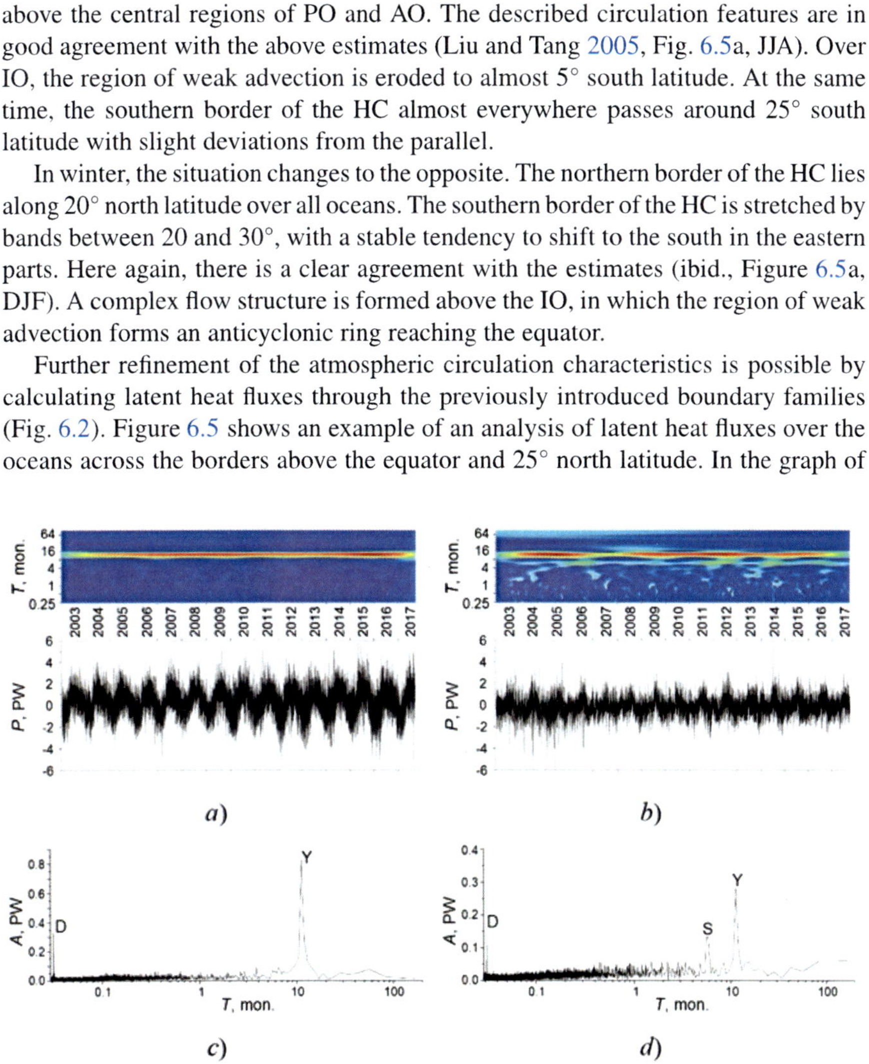

Fig. 6.5 Analysis of meridional fluxes of latent heat over the oceans: at the latitude of the equator (left); at 25° north latitude (right): time series of the total power *P*, PW of the flows with a sampling of 6 h (black curves) and three-month averaging (red curves); wavelet spectra (**a**, **b** above) and Fourier spectra (**c**, **d**) of the initial time series (explanations in the text)

Fig. 6.5a, the black line shows the calculated time series of the total latent heat flux power through the equator *P*, PW (with a sampling time of 6 h), and the red one is the result of smoothing it in a sliding window corresponding to the three-month averaging interval. Positive values of flows correspond to the direction to the north. It can be seen that a one-year period of oscillations is clearly traced in the evolution of the flow. In this case, however, the variance of deviations from the annual harmonic is much higher than its power, which indicates a significant influence on the overall circulation of processes of relatively small (synoptic and mesoscale) scales. The variability of the annual harmonic itself in the considered 15-year interval is also noticeable.

The study of the fine features of the evolution of time series is possible using the apparatus of wavelet analysis. Above the graph of Fig. 6.5a, a wavelet diagram (the result of decomposition using Morlet wavelet) of the initial time series is given. The vertical axis represents the effective scale, reduced to the time interval in months. Small scales (starting from a week, 0.25 months) are located below for the convenience of correlation with time series graphs. The horizontal time scales of wavelet and Fourier spectra are combined. The maximum scale is limited in the wavelet analysis to 96 months (8 years). The wavelet spectrum confirms that the most striking feature of the time series of the total power flow through the equator over the World Ocean is its annual fluctuation (seasonal variability). This is consistent with the correct, close to sinusoidal, shape of the smoothed time series on the graph. The picture of wavelet decomposition does not contradict the assumption of the presence of long-period (quasi-biannual, quasi-four-year, etc.) cycles, but the initial 15-year interval does not provide a reliable basis for analyzing oscillations of such scales. One can also note a rather complex, but relatively weakly expressed structure of perturbations of smaller scales (from semi-annual to weekly).

The best resolution in the frequency of harmonic oscillations (without the possibility of localization in time) is given by the Fourier analysis. The Fourier spectrum for the original series in the graph of Fig. 6.5a is shown below it in Fig. 6.5c. The horizontal coordinate is converted to the inverse frequency, i.e. period of fluctuations, measured in months. The Fourier analysis finally proves that in the evolution of latent heat fluxes above the equator, the most pronounced are the annual and daily harmonics, indicated on the graph in Fig. 6.5 by the letters "Y" and "D", respectively. This result is also in complete agreement with the known properties of atmospheric circulation.

The second example is illustrated in Figs. 6.5b, d. Here, as the initial time series, the total power of the stream across the boundary of 25° north latitude was taken (black line in the graph of Fig. 6.5c). It can be noted that the smoothed row (red line) in this case is characterized by a more irregular, rugged shape. The dispersion of deviations of the initial series from the smoothed one is comparable to that observed in flows through the equator, however, the annual harmonic power is several times less (Fig. 6.5d), which indicates the increasing contribution of small-scale processes. This is confirmed by the wavelet analysis (Fig. 6.5b above), which distinguishes not only a pronounced annual oscillation, but also a noticeable semiannual oscillation, which passes on a smaller scale into a complex bifurcation structure. The described picture is in qualitative agreement with the results of the Fourier analysis, which again reveals the presence of daily ("D") and annual ("Y") harmonics, as well as a weaker semi-annual ("S"), Fig. 6.5d. The absence of pronounced harmonics of other periods, simultaneously with a significant dispersion of the initial time series, suggests that a significant part of the transferred energy is distributed almost evenly over a large set (spectrum) of frequencies corresponding to time intervals from units of a day to units of months. This fact, probably, should also be interpreted as a manifestation of the significant contribution of the turbulent mechanism of latent heat transfer, which increases when shifted from the equator to mid-latitudes. However, as noted above, its accurate analysis requires better spatiotemporal detailing of the source data.

Figure 6.6 displays some of the results of a similar analysis performed for zonal fluxes of latent heat across the boundaries indicated in Fig. 6.2 by the letters "I" and "A". The border "I" passes over the IO along the meridian of 90° east longitude between 15° north and south latitude. The initial time series of the total latent heat flux with a step of 6 h is shown in the graph of Fig. 6.6a by a black line, smoothed in a three-month interval by a red line. Positive values of flows correspond to the eastward direction. It can be seen that the flow variations are complex, nevertheless, the annual harmonic dominates in them, which stands out well in the waveletogram of the original series shown at the top of Fig. 6.6a, and in the Fourier spectrum of the same series shown in Fig. 6.6c below. Also, irregular disturbances with scales from 1 to 4 months are observed in the waveletogram, which may be associated with Madden—Julian oscillations (MJO) (Madden and Julian 1994). It should be noted,

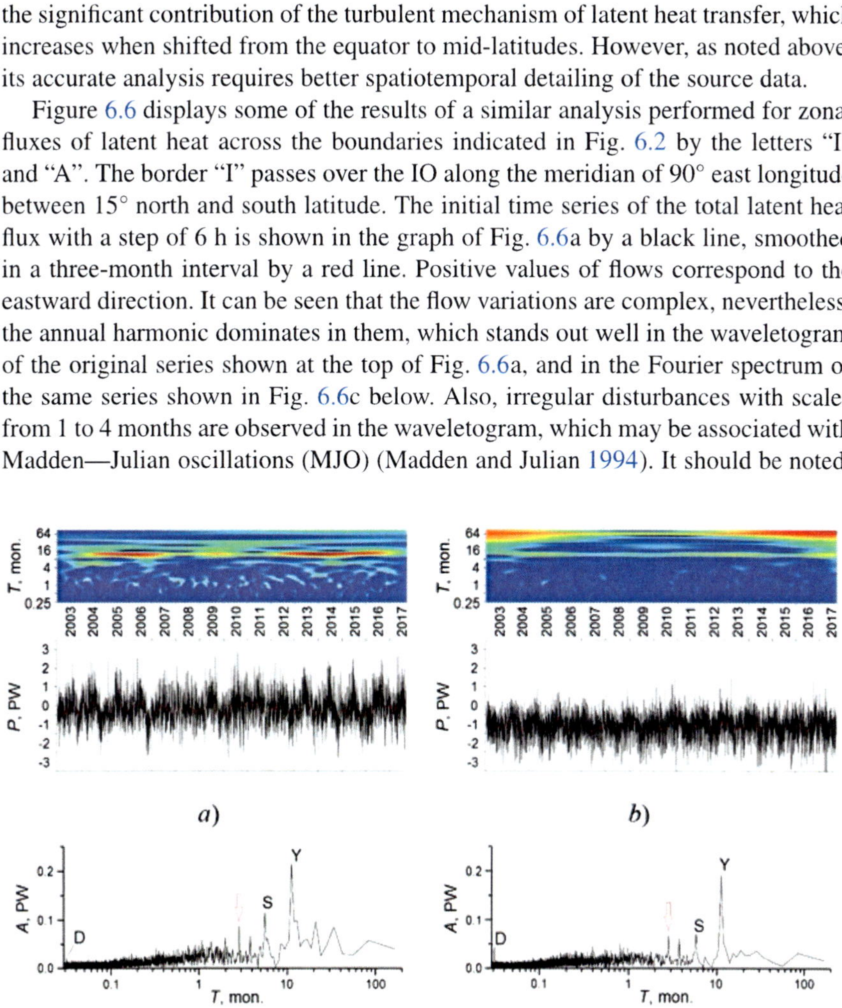

Fig. 6.6 Analysis of zonal fluxes of latent heat in the tropics over the Indian (left) and Atlantic (right) oceans: time series of the total power *P*, PW of flows with a sampling of 6 h (black curves) and three-month averaging (red curves); wavelet spectra (**a**, **b** above) and Fourier spectra (**c**, **d**) of the initial time series (explanations in the text)

however, that in the Fourier spectrum there are no pronounced harmonics with a period of 40-50 days (about 1.5 months), the closest harmonics, with a period of slightly less than 3 months, are shown in Fig. 6.6c by a red arrow. On the other hand, given the irregular formation of the Madden-Julian wave, the absence of a clearly defined spectral component in the series under study does not prove the absence of manifestations of MJO in it. This issue should be investigated in more detail, probably with the use of data of higher detail.

Figures 6.6b and 6.6d illustrate the results of a similar analysis for the border "A", which passes over the AO along the meridian of 30° east longitude between 15° north and south latitude. The eastern transport of latent heat above the IO is much more pronounced than above the IO—the flow values are on average substantially negative. As in other cases, the annual, semi-annual, and daily harmonic oscillations are clearly distinguished (see Fig. 6.6d). Perturbations with a scale of 1 to 4 months are noticeably weaker than above the IO (see the wavelet diagram in Fig. 6.6b). Analysis at other boundaries as a whole shows a monotonic attenuation of these perturbations when shifting to the east, which is also characteristic of MJO. The Fourier spectrum again shows pronounced harmonics with periods of about 3 and 4 months (but not 1.5 months).

In conclusion, the structure of meridional latent heat fluxes as functions of latitude was studied. For this purpose, the average values of the power of flows through all latitudinal boundaries, shown in Fig. 6.2, were calculated for the "summers" (warm boreal seasons) and "winters" (cold boreal seasons) of 2003–2017. To be able to directly compare them, the calculated average flux values were normalized to the lengths of the corresponding boundaries, as a result of which the specific power of the latent heat fluxes in MW/m was obtained as a function of latitude. As noted above, specific flows can be calculated separately for PO, AO and IO, and for the entire World Ocean (WO) as a whole. Figure 6.7 shows the results of the calculation of specific flows over the WO. Positive values of flows correspond to the direction of heat transfer to the north; positive values of latitudes (horizontal) to the northern hemisphere. The red curves show the latitudinal course of the flows for the "summer" of different years, and blue for the "winter". The thick black line shows the latitudinal course of the average summer flow for all years; thick blue—average winter. Diagonally shaded areas show the rms scatter of values for all years for summer (yellow hatching) and winter (gray hatching). The latitudinal variations in the specific latent heat fluxes for the summer of 2017 (dashed black line) and the winter of 2016/2017 (dashed blue line) are highlighted separately in the graphs.

The following general features of the obtained curves should be noted. The curves related to the "winter" and "summer" seasons form two compact groups that are well separated on the graph and are characterized by a relatively small intragroup dispersion. The overall picture is more symmetrical with respect to the equator in the "winter" season: heat fluxes up to 30° north latitude are negative (directed south), and vice versa, flows in the southern tropical zone are directed north, i.e. in both cases, to the equator, forming the ICZ. The group of curves corresponding to the "summer" flows on the graph runs parallel to the "winter" group, but to the right and above it. Such a shift could be the result of the addition of symmetric circulation in the

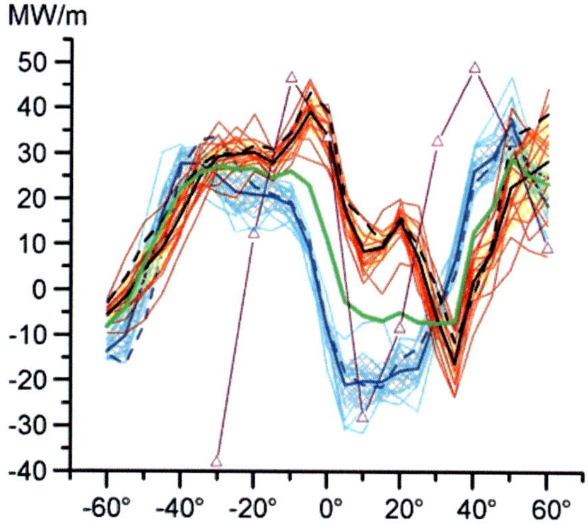

Fig. 6.7 Specific power of meridional latent heat fluxes as a function of latitude: blue curves—average "winter" flows, red curves—average "summer" flows; dashed blue curve—average flow of winter 2016/2017; dashed black curve—average flow of summer 2017; the thick blue curve is the average "winter" stream for all years, the thick black curve is the average "summer" stream; thick green curve—average flux for the continuous observation interval 2003–2016; triangles—recalculation of data (Palmén and Newton 1969) (explanations in the text)

Headley cells with some additional process, which carries out the directed transfer of latent heat from the southern hemisphere to the northern one. In (Palmén and Newton 1969) it is indicated that on average condensation prevails over evaporation in the northern hemisphere, and evaporation over condensation prevails over southern. To compensate for this difference, a flow of water vapor from the southern hemisphere to the northern hemisphere is needed, corresponding to the transfer of about 1.647 × 1016 kg of water per year (or about 1.3 PW of latent heat flow power). Based on the constructed graphs, an assumption can be made that this process is intensifying in the "summer" season. It should also be noted that the latitudinal shift between the groups of "winter" and "summer" flows reflects well the seasonal migration of ICZ.

Additionally, the latitudinal variation of the specific powers of the latent heat fluxes averaged over the entire observation period, excluding the incomplete 2017, is plotted on the graph in Fig. 6.7 (green curve). The intersection of this curve with the level of zero flows falls at 5° north latitude, which is in good agreement with the average position of the thermal equator (Palmén and Newton 1969), and the average value of the flux through the geographic equator is positive, which corresponds to the transfer of latent heat to the north. Subtracting this positive "additive" of the order of 11 MW/m, the characteristic symmetric structure of flows directed to the equator in tropical latitudes is also clearly visible. It should be noted that the half-width of the HC, estimated from considerations of "symmetry" of the constructed curve of average flows, is much larger than expected—about 40° in each hemisphere. One of the possible reasons for this result is discussed above and concludes that small-scale turbulent processes, the analysis of which requires the construction of data series of higher spatiotemporal resolution, can be an important factor in the latent heat advection. On the other hand, studies of the beginning of the century suggest that the trend of "widening of the tropics" is revealed in numerous series of independent satellite observations (Reichler 2009; Lu et al. 2009), capable of affecting the width of the HC in the direction of some increase, see also (Kossin et al. 2014; Pan et al. 2017).

To quantitatively compare the obtained mean fluxes, the graph in Fig. 6.7 also shows a calculation based on data from (Palmén and Newton 1969), where model estimates of the total latent heat fluxes through latitudinal boundaries were performed with a step of 10° (ibid., Table 2.4, p. 39). These values can easily be translated into estimates of specific flows, dividing by the lengths of the corresponding parallels. The obtained values are shown in Fig. 6.7 by triangles connected by a purple line. Between 60° north latitude and 20° south latitude, the results of both calculations, in order of magnitude, are in good agreement with each other (south of this region and up to 60° south latitude (Palmén and Newton 1969) give much larger estimates of flows in the south). Moreover, as already noted, the features of the latitudinal course of the average flows calculated by the satellite radiothermovision algorithms and those given in (Palmén and Newton 1969) differ markedly.

It should also be noted that the nature of the latitudinal dependence of winter and summer flows in the low latitude belt qualitatively corresponds to the dependence obtained in (Liu and Tang 2005, Fig. 6.7a, plots). Some differences can be caused by the fact that the time boundaries of the seasons in the works are chosen somewhat differently, and the seasonal course of the meridional flows is significant (Liu and Tang 2005, Fig. 6.6b). This, in part, may cause substantial differences in the absolute values of the flows. It should be emphasized that, despite the consolidation of efforts to eliminate the existing uncertainties in the climatological data series, the discrepancies in the estimates of atmospheric moisture fluxes by different methods remain very significant (Robertson et al. 2014). So, the graphs in (Liu and Tang 2005, Fig. 6.7a) allow, for example, to estimate the specific power of the average annual meridional latent heat flux through the equator, as the average between the "winter" and "summer" fluxes. According to calculations (Liu and Tang 2005), this value should be more than 100 MW/m (over the ocean); according to climatological estimates of J.P. Peixoto and A. Oort the same value is approximately 23 MW/m (average above the equator) (Peixoto and Oort 1983); calculations using satellite radiothermovision algorithms give a value of about 11 MW/m; estimation based on the data presented in (Palmén and Newton 1969) gives about 33 MW/m, Fig. 6.7.

The graphs of the average annual specific power of the latent heat flow over the World Ocean (as a function of latitude) for different years of observations form a more compact structure, shown in Fig. 6.8.

Moreover, all the main features of the latitudinal dependence of the flows noted above regarding asymmetry with respect to the equator are also manifested in it. Table 6.1 summarizes the results of calculations of the characteristics of the meridional circulation for the entire oceans: for a given latitude θ, the length of the corresponding boundary L is listed, along with the total power of the latent heat flux

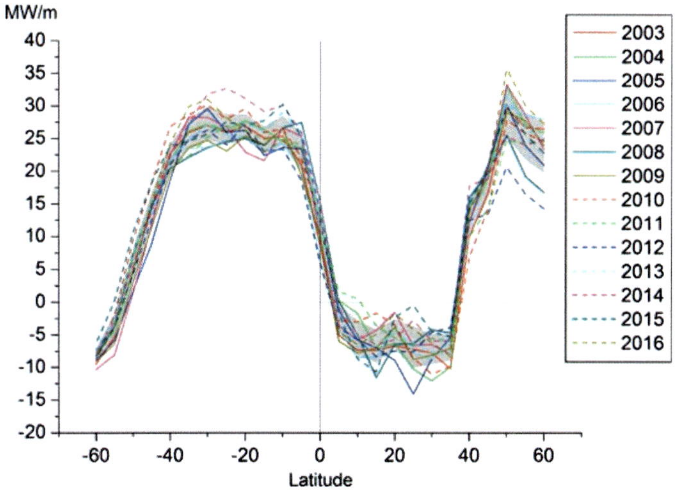

Fig. 6.8 Specific power (MW/m) of average annual latent heat fluxes over the World Ocean through latitudinal boundaries from 60° S up to 60° N in increments of 5° for 2003–2016. Positive latitudes (along the abscissa) are for the northern hemisphere. Positive values of flows (along the ordinate axis)—for the direction to the north. The gray bar is the range of specific power values (for a given latitude) that differ from the average for all years by no more than one standard deviation

through the boundary Q, the average specific power P for all years, the standard deviation of the average annual specific flux σ_P, the minimum $P_{\min}$ and maximum $P_{\max}$ values of the average annual specific flow rate for all years (in parentheses is the year the corresponding value was reached).

In the adopted calculation procedure, a percentage of the area of the ocean and land in different latitudinal zones, as well as their distribution between the southern and northern hemispheres, can have a certain influence on the type of flow diagrams. In this regard, of interest are the results of calculations of the average annual specific meridional latent heat fluxes for the Pacific, Atlantic and Indian oceans separately. Thus, the Pacific Ocean covers an area of about half of the World Ocean and is located almost symmetrically relative to the equator. The calculation results for the Pacific Ocean are shown in Fig. 6.9.

The main features of the obtained flow structure are briefly summarized below.

(1) There is some tendency to increase the symmetry of the flows relative to the equator due to a significant (about one and a half times) decrease in flows in the tropical latitudes of the southern hemisphere.
(2) The average level of specific flow through the equator is halved to 5.7 MW/m, but remains positive (see Table 6.2). The total flow through the equator with a border length of 15817 km is 0.09 PW (about 27% of the total).
(3) The intersection of the abscissa axis, as in Fig. 6.8, occurs at 5° N, which corresponds well to the average position of the thermal equator. It should be additionally noted that the position of this point is most stable on the graphs of

Table 6.1 Meridional circulation of latent heat over the World Ocean

θ, °	L, km	Q, PW	P, MW/m	σ_P, MW/m	P_{min}, MW/m	P_{max}, MW/m
60	5310	0.12	23.5	3.9	14.3 (2012)	27.9 (2006)
55	8770	0.22	25.3	3.7	16.5 (2012)	29.9 (2016)
50	10,346	0.30	29.0	4.0	20.6 (2012)	35.7 (2016)
45	13,779	0.25	18.3	2.6	13.5 (2012)	21.0 (2005)
40	16,163	0.21	12.7	1.8	6.8 (2010)	17.8 (2014)
35	18,240	−0.13	−7.0	2.1	−10.3 (2003)	−3.9 (2012)
30	18,971	−0.14	−7.3	2.4	−12.1 (2004)	−4.2 (2008)
25	21,390	−0.15	−7.0	3.2	−14.1 (2005)	−0.5 (2015)
20	24,486	−0.12	−4.9	2.2	−8.9 (2005)	−1.6 (2007)
15	27,979	−0.19	−6.8	2.6	−11.6 (2008)	−1.7 (2010)
10	28,636	−0.16	−5.5	2.6	−8.7 (2015)	0.5 (2011)
5	29,825	−0.07	−2.5	2.3	−6.0 (2009)	1.6 (2011)
0	29,856	0.33	11.0	2.9	5.8 (2012)	15.3 (2008)
−5	29,244	0.67	23.1	2.7	18.5 (2012)	27.4 (2008)
−10	30,433	0.80	26.2	2.2	23.4 (2012)	30.3 (2015)
−15	29,295	0.72	24.7	2.1	21.5 (2007)	29.1 (2014)
−20	27,899	0.74	26.7	2.0	22.9 (2007)	31.1 (2014)
−25	27,487	0.73	26.4	2.3	23.0 (2009)	32.6 (2014)
−30	27,541	0.75	27.2	2.6	23.6 (2008)	31.7 (2014)
−35	29,512	0.76	25.6	2.4	22.2 (2008)	29.9 (2016)
−40	29,196	0.64	22.1	2.0	18.6 (2005)	26.8 (2010)
−45	27,303	0.39	14.2	2.5	8.9 (2005)	18.4 (2015)
−50	25,141	0.14	5.5	2.7	1.3 (2007)	10.7 (2015)
−55	22,642	−0.09	−4.0	2.1	−8.1 (2007)	0.2 (2015)
−60	18,264	−0.15	−8.3	1.0	−10.4 (2007)	−6.4 (2015)

different years for flows over the World Ocean (Fig. 6.8) and, to a lesser extent, for flows over the Pacific Ocean.

(4) The width of the HC looks almost the same in Figs. 6.8 and 6.9.

The calculation results are shown in Table 6.2.

The results of similar calculations for the Atlantic Ocean are shown in Fig. 6.10 and are summarized in Table 6.3.

Fluxes over the Atlantic are on average more intense than over the Pacific Ocean, and, in accordance with existing conceptions, show significant asymmetry relative to the equator:

(1) The maximum specific fluxes in the southern hemisphere can exceed 50 MW/m, while in the northern hemisphere reach 20 MW/m in the HC and 47 MW/m in high latitudes.

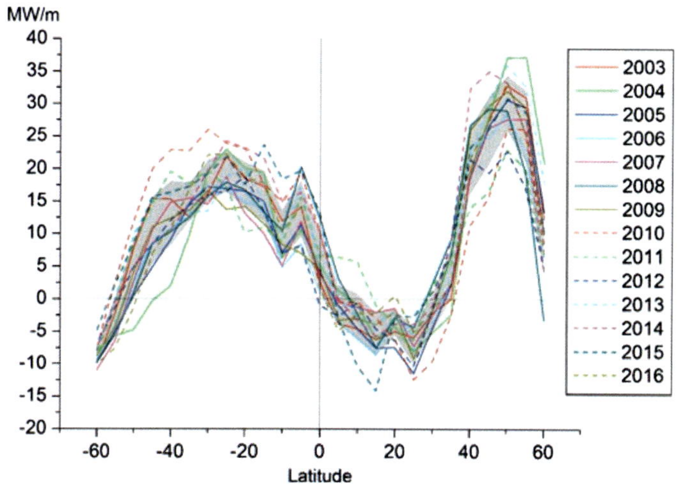

Fig. 6.9 Specific power (MW/m) of average annual latent heat fluxes over the Pacific Ocean through latitudinal boundaries from 60° S up to 60° N in increments of 5° for 2003–2016. Positive latitudes (along the abscissa) are for the northern hemisphere. Positive values of flows (along the ordinate axis)—for the direction to the north. The gray bar is the range of specific power values (for a given latitude) that differ from the average for all years by no more than one standard deviation

(2) The average level of specific flow through the equator is 25 MW/m. The total flow through the equator with a border length of 6144 km is 0.15 PW (about 46% of the total).
(3) The intersection of the abscissa axis shifts to approximately 10° N
(4) The width of the HC is slightly reduced compared to the Pacific.

In conclusion, the results of calculations over the Indian Ocean are considered, Fig. 6.11.

It is clear that the general asymmetry of flows relative to the equator is significantly affected by the predominance of land in the northern hemisphere. Nevertheless, in the latitude region available for analysis, the intensity of latent heat fluxes over the Indian Ocean is generally higher than over the Pacific Ocean, and in some places approaches that over the Atlantic:

(1) The maximum specific flows in the southern hemisphere exceed 50 MW/m, in the northern—30 MW/m.
(2) The average flow rate through the equator is 10.5 MW/m. The total flow through the equator with a border length of 6088 km is 0.06 PW (about 20% of the total). As a result, the sum of the total flows over three oceans through the equator gives 0.31 PW, or 93% of the total flux previously calculated over the World Ocean. The remaining 7%, or about 0.02 PW, falls on excluded border fragments with a total length of about 1807 km, mainly over the inland seas of the Indonesian archipelago and near coastlines. Thus, the average specific flux through these fragments of the boundaries is less than 12 MW/m, which

Table 6.2 Meridional circulation of latent heat over the Pacific Ocean

θ, °	L, km	Q, PW	P, MW/m	σ_P, MW/m	P_{min}, MW/m	P_{max}, MW/m
60	1154	0.01	9.0	6.0	−3.1 (2008)	20.9 (2004)
55	3699	0.10	26.5	5.5	17.0 (2012)	37.3 (2004)
50	6111	0.18	30.2	4.4	22.8 (2012)	37.2 (2004)
45	7253	0.19	25.9	5.4	15.6 (2010)	35.0 (2014)
40	8837	0.19	21.2	5.3	11.2 (2010)	32.4 (2014)
35	10,019	0.04	4.1	4.1	−2.6 (2010)	9.2 (2014)
30	11,435	−0.02	−2.1	3.1	−9.6 (2010)	2.3 (2008)
25	12,496	−0.09	−7.2	2.9	−12.5 (2010)	−2.7 (2015)
20	14,707	−0.05	−3.6	2.2	−7.4 (2005)	0.5 (2016)
15	16,567	−0.09	−5.4	3.4	−14.1 (2015)	−1.0 (2011)
10	17,630	−0.05	−2.6	3.5	−10.3 (2015)	5.7 (2011)
5	18,637	−0.02	−0.9	3.1	−4.9 (2016)	6.5 (2011)
0	15,817	0.09	5.7	4.3	−1.0 (2012)	13.2 (2015)
−5	13,985	0.19	13.9	4.4	7.0 (2009)	20.3 (2008)
−10	13,716	0.13	9.8	4.0	4.8 (2006)	18.4 (2015)
−15	14,392	0.23	15.8	4.0	9.9 (2007)	23.7 (2015)
−20	13,401	0.23	17.5	3.6	10.2 (2011)	23.1 (2014)
−25	13,529	0.26	19.6	3.2	13.7 (2009)	24.2 (2014)
−30	12,904	0.23	18.1	3.3	13.4 (2013)	26.0 (2010)
−35	12,251	0.18	14.8	3.1	9.7 (2004)	22.7 (2010)
−40	11,372	0.15	13.0	5.1	1.9 (2004)	22.9 (2010)
−45	10,497	0.11	10.5	5.3	−0.3 (2004)	19.9 (2010)
−50	9703	0.04	4.4	4.7	−4.8 (2004)	12.0 (2010)
−55	9025	−0.02	−2.7	3.2	−7.5 (2016)	2.6 (2010)
−60	8354	−0.07	−8.6	1.6	−11.1 (2007)	−4.9 (2015)

approximately corresponds to the average specific flux through the equator over the entire oceans (11 MW/m). Nevertheless, it should be noted that the estimates of advection over such areas are the least reliable and can be one of the significant sources of errors.

(3) The intersection of the abscissa axis occurs at about 5° N.
(4) The width and configuration of the HC over the Indian Ocean is a complex issue. In the northern hemisphere, it can partially capture the land area. In the southern hemisphere, there are two pronounced maximums of flows about 10 and 35° S, which are not observed in the structure of meridional flows over other oceans.

The results of calculation over Indian Ocean are shown in Table 6.4.

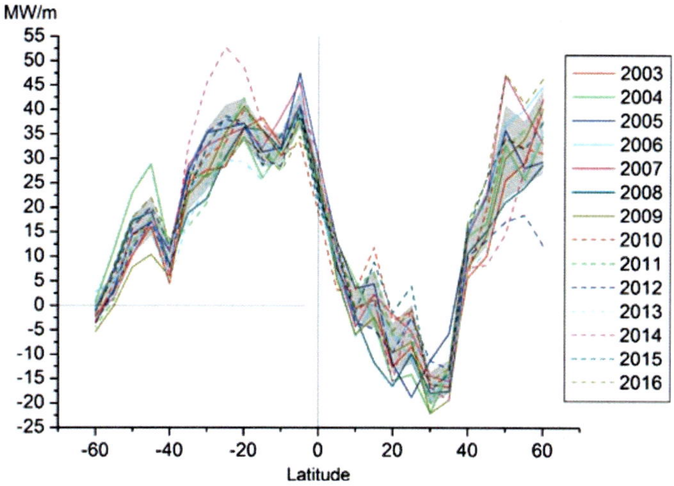

Fig. 6.10 Specific power (MW/m) of average annual latent heat fluxes over the Atlantic Ocean through latitudinal boundaries from 60° S up to 60° N in increments of 5° for 2003–2016, see the notes to Fig. 6.9

6.4 Concluding Remarks to this Chapter

1. The results presented in this Chapter relate to studies of global and regional (over individual oceans) atmospheric circulation at climatically significant time intervals.
2. The approach of satellite radiothermovision allowed computing a detailed (0.25° grid with a time step of 6 h) picture of global atmospheric circulation exclusively on the basis of long-term observational data.
3. The calculations realistically reproduce many of characteristic elements and parameters of global atmosphere, such as: the zonal structure, the boundaries of the circulation cells, the predominance of zonal transport over the meridional one, and the characteristic velocities and directions of advection in the lower troposphere over all basins of the World Ocean; annual and seasonal variations, manifested in the migration of the intertropical convergence zone and the boundaries of the Hadlye cells; and harmonics of the latent heat fluxes with periods of 0.033 months (day), 6 and 12 months; the average position of the thermal equator at 5° north latitude, the convergence of latent heat fluxes in the equatorial zone, the average positive latent heat flux from the southern hemisphere to the northern. Thus, it is shown that *information on all the indicated properties of atmospheric circulation is directly contained in satellite observation data and can be effectively extracted from them using the satellite radiothermovision approach.*

Table 6.3 Meridional circulation of latent heat over the Atlantic Ocean

θ, °	L, km	Q, PW	P, MW/m	σ_P, MW/m	P_{min}, MW/m	P_{max}, MW/m
60	2418	0.09	35.5	8.7	12.3 (2012)	46.4 (2016)
55	3699	0.11	31.1	6.6	18.5 (2012)	41.1 (2016)
50	3556	0.11	31.2	9.7	14.5 (2014)	47.6 (2016)
45	4639	0.08	17.7	5.8	8.2 (2014)	26.0 (2015)
40	5239	0.06	11.6	4.2	5.6 (2003)	17.2 (2016)
35	6217	−0.09	−14.6	3.4	−19.5 (2014)	−5.7 (2005)
30	6645	−0.11	−17.0	3.3	−22.1 (2009)	−11.2 (2005)
25	7684	−0.05	−6.0	6.1	−18.8 (2005)	4.0 (2015)
20	5695	−0.05	−8.8	5.1	−16.6 (2008)	−1.5 (2015)
15	6605	0.01	1.3	5.9	−11.6 (2008)	11.8 (2010)
10	4983	−0.00	−0.8	3.6	−6.4 (2006)	5.0 (2004)
5	4625	0.04	9.1	3.1	3.2 (2010)	13.1 (2005)
0	6144	0.15	24.7	3.4	18.8 (2010)	30.6 (2014)
−5	5206	0.21	39.8	3.8	33.2 (2010)	47.5 (2005)
−10	5311	0.17	32.0	3.0	27.4 (2016)	38.5 (2007)
−15	5343	0.17	32.2	4.0	25.5 (2013)	38.3 (2003)
−20	5355	0.20	37.8	4.5	29.3 (2013)	48.7 (2014)
−25	6147	0.21	34.8	6.2	28.3 (2003)	52.7 (2014)
−30	6283	0.19	29.8	6.4	20.9 (2011)	45.5 (2014)
−35	6558	0.16	23.9	4.5	16.0 (2011)	32.8 (2014)
−40	6793	0.06	9.0	3.3	4.3 (2003)	13.3 (2016)
−45	6506	0.12	18.1	4.1	10.4 (2009)	28.8 (2004)
−50	6075	0.08	14.0	3.9	7.8 (2009)	23.0 (2004)
−55	5134	0.02	4.7	3.1	−0.1 (2009)	11.3 (2004)
−60	2752	−0.01	−2.0	2.1	−5.5 (2009)	2.8 (2006)

4. The latitudinal distributions of annual and seasonal meridional fluxes of latent heat over the World Ocean as a whole, as well as the Pacific, Atlantic and Indian oceans are constructed. The obtained distributions are in qualitative agreement with the known modeling data, and perhaps manifest the climatic tendency noted in a number of works as "widening of the tropics".

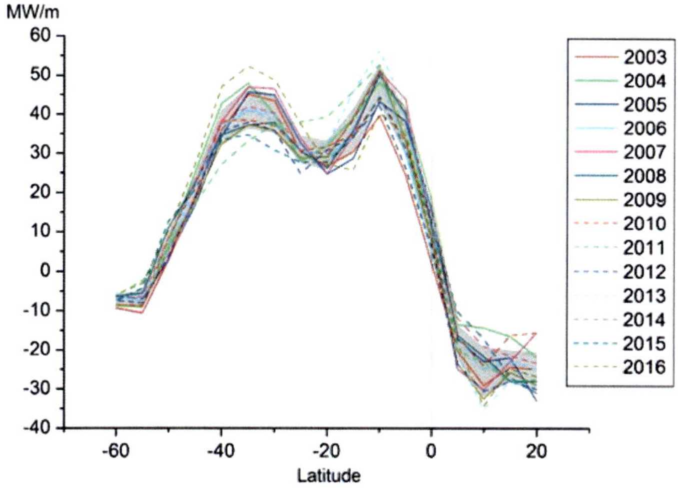

Fig. 6.11 Specific power (MW/m) of average annual latent heat fluxes over the Indian Ocean through latitudinal boundaries from 60° S up to 20° N in increments of 5° for 2003–2016, see the notes to Fig. 6.9

Table 6.4 Meridional circulation of latent heat over the Indian Ocean

θ, °	L, km	Q, PW	P, MW/m	σ_P, MW/m	$P_{\min}$, MW/m	$P_{\max}$, MW/m
20	2116	−0.05	−25.8	5.3	−33.0 (2005)	−15.5 (2007)
15	4243	−0.10	−24.1	4.0	−28.2 (2013)	−16.2 (2010)
10	4599	−0.12	−25.7	6.5	−35.3 (2013)	−14.3 (2004)
5	5123	−0.09	−18.1	4.4	−25.0 (2007)	−10.2 (2015)
0	6088	0.06	10.5	4.6	2.1 (2003)	18.2 (2004)
−5	6951	0.25	35.3	5.9	24.5 (2003)	43.7 (2007)
−10	8596	0.41	47.5	5.1	39.6 (2003)	56.3 (2013)
−15	7894	0.28	35.8	5.7	25.7 (2016)	45.7 (2011)
−20	6896	0.20	29.4	3.8	24.7 (2005)	39.3 (2011)
−25	7533	0.23	31.0	3.6	24.8 (2012)	38.0 (2011)
−30	7969	0.32	39.8	4.8	30.8 (2015)	48.7 (2016)
−35	8744	0.36	41.5	5.5	33.2 (2011)	52.3 (2016)
−40	8284	0.30	36.4	4.7	27.2 (2011)	47.0 (2016)
−45	7646	0.16	20.8	3.1	16.7 (2005)	27.2 (2016)
−50	6951	0.05	6.6	3.1	2.6 (2007)	12.7 (2015)
−55	6202	−0.04	−7.1	2.6	−10.7 (2007)	−2.4 (2011)
−60	5407	−0.04	−7.4	1.3	−9.4 (2007)	−5.4 (2006)

References

Barry RG, Chorley RJ (2003) Atmosphere, weather and climate. 8th edn. Routledge, Taylor & Francis Group, London and New York, 420 p

Budyko MI (1956) Teplovoy balans zemnnoy poverkhnosti. Leningrad: Gidrometeorologicheskoye Izdatel'stvo, 255 p. (English transl., Stepanova NA, The Heat Balance of the Earth's Surface. Office Tech. Serv., U. S. Dept. Commerce, Washington D.C., 1958)

Budyko MI (1963) Atlas Teplovogo Balansa Zemnogo Shara (Atlas of heat balance of the globe). USSR, Moscow: GGO, 69 p

Emanuel K (2003) Tropical cyclones. Annu Rev Earth Planet Sci 31:75–104

Ermakov DM (2017) Investigation of the features of long-term global atmospheric circulation via satellite radiothermovision. In: Proceedings of 2017 progress in electromagnetics research symposium—spring (PIERS), pp 413–418. https://doi.org/10.1109/piers.2017.8261775

Ermakov DM (2018) Global circulation of latent heat in the Earth's atmosphere according to data from satellite radiothermovision. Izvestiya, Atmos Oceanic Phys 54(9):1223–1243

Ermakov D, Chernushich A, Sharkov E, Shramkov Y (2011) Stream Handler system: an experience of application to investigation of global tropical cyclogenesis (electronic resource). In: Proceedings of 34th international symposium on remote sensing of environment, Sydney, pp 10–15 Apr, 2011. 2011. http://www.isprs.org/proceedings/2011/ISRSE-34/211104015Final00456.pdf

Ermakov DM, Sharkov EA Chernushich AP (2015) Satellite radiothermovision of atmospheric mesoscale processes: case study of tropical cyclones. In: The international archives of the photogrammetry, remote sensing and spatial information sciences—ISPRS Archives. vol XL, no 7/W3. pp 179–186

Ermakov DM, Chernushich AP, Sharkov EA (2016a). Geoportal of satellite radiothermovision: data, services, prospects. Sovremennye Problemy Distantsionnogo Zondirovaniya Zemli iz Kosmosa (Current Problems of Remote Sensing of Earth from Space) 13(3):46–57. https://doi.org/10.21046/2070-7401-2016-13-3-46-57 (in Russian)

Ermakov DM, Sharkov EA, Chernushich AP (2016b) A multisensory algorithm of satellite radiothermovision. Izvestiya, Atmos Oceanic Phys 52(9):1172–1180

Ermakov DM, Sharkov EA, Chernushich AP (2017a) Satellite radiothermovision on synoptic and climatically significant scales. Izvestiya, Atmos Oceanic Phys 53(9):973–978

Ermakov DM, Sharkov EA, Chernushich AP (2017b) Satellite radiothermovision analysis of the evolution of a system of interacting typhoons. Izvestiya, Atmos Oceanic Phys 53(9):945–954

Ermakov DM, Raev MD, Chernushich AP, Sharkov EA (2019a) Algorithm for construction of global ocean-atmosphere radiothermal fields with high spatiotemporal sampling based on satellite microwave measurements. Izvestiya, Atmos Oceanic Phys 55(9):1041–1052. https://doi.org/10.1134/s0001433819090159

Ermakov DM, Sharkov EA, Chernushich AP (2019b) Evaluation of tropospheric latent heat advective fluxes over the ocean by the animated analysis of satellite radiothermal remote data. Izvestiya, Atmos Oceanic Phys 55(9):1125–1132. https://doi.org/10.1134/s0001433819090160

Eyring V, Bony S, Meehl GA, Senior C, Stevens B, Stouffer RJ, Taylor KE (2015) Overview of the coupled model intercomparison project phase 6 (CMIP6) experimental design and organization. Geosci Model Dev Discuss 8(12):10539–10583

Feulner G, Rahmstorf S, Levermann A, Volkwardt S (2013) On the origin of the surface air temperature difference between the hemispheres in Earth's present-day climate. J Clim 26(18):7136–7150

Golitsyn GS (2008) Polar lows and tropical hurricanes: their energy and sizes and a quantitative criterion for their generation. Izvestiya, Atmos Oceanic Phys 44(5):537–547. https://doi.org/10.1134/S0001433808050010

Haltiner GJ, Martin FL (1957) Dynamical and physical meteorology. McGraw-Hill Book Company, Inc., New York, Toronto, London 1957. 470 p

Henderson-Sellers A, McGuffie K (2012) The future of the World's climate. Elsevier, Amsterdam, ND, p 660

Kang SM, Seager R (2015) Croll revisited: why is the northern hemisphere warmer than the southern hemisphere? Clim Dyn 44(4–5):1457–1472

Kossin JP, Emanuel KA, Vecchi GA (2014) The poleward migration of the location of tropical cyclone maximum intensity. Nature 509:349–352

Liu WT, Tang W (2005) Estimating moisture transport over oceans using space-based observations. J Geophys Res 110(D10):D10101. https://doi.org/10.1029/2004JD005300

Lu J, Deser C, Reichler T (2009) Cause of the widening of the tropical belt since 1958. Geophys Res Lett 36:L03803. https://doi.org/10.1029/2008gl036076

Madden R.A., Julian P.R (1994) Observations of the 40-50-day tropical oscillation. Mon Weather Rev 122(5):814–837

Newell RE, Newell NE, Zhu Y, Scott C (1992) Tropospheric rivers?—a pilot study. Geophys Res Lett 19(24):2401–2404

Palmén E, Newton CW (1969) Atmospheric circulation systems: their structural and physical interpretation. Academic Press. New York, 1969. XVIII + 606 p

Pan Y, Li L, Jiang X, Li G, Zhang W, Wang X, Ingersol AP (2017) Earth's changing global atmospheric energy cycle in response to climate change. Nature Commun 8(14367). https://doi.org/10.1038/ncomms14367

Peixoto JP, Oort AH (1983) The atmospheric branch of the hydrological cycle and climate. In: Street-Perrott A, Beran M, Ratcliffe R, Variations in the global water budget. Springer, New York, pp 5–65

Reichler T (2009) Changes in the atmospheric circulation as indicator of climate change. In: Climate change: observed impacts on planet earth (Ed. Trevor M. Letcher). Elsevier, pp 145–164

Robertson FR, Bosilovich MG, Roberts JB, Reichle RH, Adler R, Ricciardully L, Berg W, Huffman GJ (2014) Consistency of estimated global water cycle variations over the satellite era. J Clim 27(16):6135–6154

Sellers WD (1965) Physical climatology. University of Chicago Press, Chicago. 272 p

Sharkov EA (2000) Global tropical cyclogenesis. Springer/PRAXIS, Berlin, Heidelberg, London, NY etc., 361 p

Sharkov EA (2012) Global tropical cyclogenesis. 2nd edn, Springer/Praxis 603 p

Taylor KE, Souffer RJ, Meehl GA (2012) An overview of CMIP5 and the experiment design. Bull Am Meteorol Soc 93(4):485–498

Trenberth KE, Caron JM (2001) Estimates of meridional atmosphere and ocean heat transports. J Clim 2001 14(16):3433–3443

Volodin EM, Mortikov EV, Kostrykin SV, Galin VY, Lykossov VN, Gritsun AS, Diansky NA, Gusev AV, Iakovlev NG (2017) Simulation of the present-day climate with the climate model INMCM5. Clim Dyn 49:3715–3734

Wentz FJ, Hilburn KA, Smith DK (2012) Remote sensing systems DMSP SSM/I, SSMIS daily environmental suite on 0.25 deg grid, version 7, 8 (electronic resource). Remote sensing systems, Santa Rosa, CA. 2012. URL: http://www.remss.com/missions/ssmi/

Wentz FJ, Ricciardulli L, Gentemann C, Meissner T, Hilburn KA, Scott J (2013) Remote sensing systems coriolis windsat daily environmental suite on 0.25 deg grid, version 7.0.1 (electronic resource). Remote Sensing Systems, Santa Rosa, CA, 2013. http://www.remss.com/missions/windsat

Wentz FJ, Meissner T, Gentemann C, Brewer M (2014a) Remote sensing systems AQUA AMSR-E daily environmental suite on 0.25 deg grid, version 7 (electronic resource). Remote Sensing Systems, Santa Rosa, CA. 2014. http://www.remss.com/missions/amsr

Wentz FJ, Meissner T, Gentemann C, Hilburn KA, Scott J (2014b) Remote sensing systems GCOM-W1 AMSR2 daily environmental suite on 0.25 deg grid, version 7.2 (electronic resource). Remote Sensing Systems, Santa Rosa, CA. 2014. www.remss.com/missions/amsr

Wick GA, Neiman PJ, Ralph FM (2013) Description and validation of an automated objective technique for identification and characterization of the integrated water vapor signature of atmospheric rivers. IEEE Trans Geosci Remote Sens 51(4):2166–2176

Wunch C (2005) The total meridional heat flux and its oceanic and atmospheric partition. J Clim 18(21):4374–4380
Zhu Y, Newell RE (1994) Atmospheric rivers and bombs. Geophys Res Lett 21(18):1999–2002
Zhu Y, Newell RE (1998) A proposed algorithm for moisture fluxes from atmospheric rivers. Mon Weather Rev 126(3):725–735

Chapter 7
Welcome to the Geoportal of Satellite Radiothermovision

As shown in the previous chapters, the approach of satellite radiothermovision is without limitation applicable to the study of various mesoscale and synoptic processes. However, the illustration of its application is compulsorily limited to a few specific examples (tropical cyclones, atmospheric rivers, global circulation). All possible applications cannot be covered in all their diversity. In this regard, the task of effective information support of a wide scientific community with the results of calculations of the dynamics of atmospheric fields over long observation intervals is becoming very urgent. This problem was solved by creating a geoportal of satellite radiothermovision. The main results of these studies are described in detail in publications (Ermakov et al. 2016a; Ermakov and Chernushich 2017a, b).

7.1 The Concept of Geoportal of Satellite Radiothermovision and the Network Service ICAR

Current trends in the development of network services related to solving the problems of remote sensing of the Earth are to ensure maximum transparency (access transparency achieved through "virtualization") of remote data and the procedures for their processing and analysis (Lee et al. 2011; Gorelick et al. 2017; Loupian et al. 2018; Yao et al. 2020). Ideally, the end user is spared the need to organize his own infrastructure for local storage and processing of data and is provided with remote software for their thematic analysis. The obvious goal is to expand the range of potential consumers of network services in the field of remote sensing and increase the efficiency of their work. The complex technological aspects of virtualization include the creation of scalable functional user interfaces for remote analysis. A possible approach has been introduced and is being developed as part of the geoportal of satellite radiothermovision and the ICAR network service.

D. M. Ermakov, *Satellite Radiothermovision of Atmospheric Processes*, Springer Praxis Books,
https://doi.org/10.1007/978-3-030-57085-9_7

It is important to emphasize that the geoportal of satellite radiothermovision (hereinafter—Geoportal) and the network service ICAR (hereinafter—ICAR) are completely autonomous from each other. Geoportal provides users with access to the results of calculations of the dynamics of geophysical fields of the atmosphere; for Geoportal, ICAR is one of the integrated tools for working with data. ICAR, as such, solves the tasks of virtualization and joint remote data processing of distributed satellite information centers; Geoportal for ICAR is one of the (priority) sources of information. Along with this, mechanisms for obtaining data from other network resources are also implemented.

Expansion of the Geoportal's products open to the user is carried out, firstly, by progressive (as satellite information arrives) and retrospective (according to archive data) calculation of interpolated fields of atmospheric parameters. To date, dynamic generation of various types of fields of global coverage in the seventeen-year interval of satellite observations has been provided (Ermakov and Chernushich 2017a). Secondly, the introduction of new types of processing products is being performed. The milestone in the development of the Geoportal was the provision of open access to the fields of the velocity of advection of water vapor in the lower troposphere. The vector nature of this data type necessitated the refinement of the software infrastructure. The vector data loading mechanism in a special binary format described on the Geoportal website (https://fireras.su/tpw/Radiothermovision.aspx) was additionally implemented.

The expansion of the range of Geoportal products and the ICAR service is closely related to the availability of network resources available for virtualization. Given the specialization of the Geoportal, of interest, first of all, are long ("climatic") series of data from global Earth observations. Spatial resolution can be quite rough—up to tens of kilometers. Prerequisites have been created for the development of virtual integration tools with several archives of remote data and products of their processing (Ermakov and Chernushich 2017a).

When implementing remote information processing in ICAR, an important role is played by the control of the availability of virtually integrated data, the violation of which can be caused by a change in the format, version, and access mode. So in 2017, there was a double need to modify the data loaders of the archive of Remote Sensing Systems (www.remss.com) in connection with the transition of the company from anonymous access mode via ftp to the authorized mode and changing the product version (ocean surface temperature) while changing the structure directory and introducing a new presentation format (netcdf). In both cases the problems were smoothly resolved due to impeccable work of RSS services. As a rule, the elimination of identified problems requires an individual approach, which leads to additional labor costs. It should be noted that this circumstance is genetically inherent in the ideology of virtual integration and can be eliminated only through close cooperation between archival centers and geoportals.

Execution of operations on virtually integrated data in ICAR is possible not only with the help of graphical, but also with the software interfaces exported by it (https://fireras.su/tpw/Software.aspx) which is the recommended approach. Calculation results, including ordering data, can be virtually integrated by third-party geoportals. So ICAR can be used by other geoportals as a virtualization gateway for distributed network resources based on a unified software interface.

7.2 Geoportal of Satellite Radiothermovision: Description of Data and Services

The relevance of the lines of research outlined above is also confirmed by the development of specialized geoportals representing products of interpolation processing of satellite data. Among them is the geoportal of the SSEC/CIMSS (Space Science and Engineering Center/Cooperative Institute for Meteorological Satellite Studies) of the University of Wisconsin, USA with the product MIMIC-TPW (2012), based on the approach (Wimmers and Velden 2011). Note that the second version of the product, MIMIC-TPW ver.2 (2020), has been recently released. The geoportal of satellite radiothermovision presented in the dissertation is different in two important aspects. First, an original interpolation processing scheme was applied, which makes it possible to expand the product range with fundamentally new types (Ermakov et al. 2017). Secondly, the geoportal architecture is a significant evolutionary step compared to the early approach (Ermakov et al. 2007) and takes into account the most current trends in this area (Savorskiy et al. 2014). It is based on the idea of dynamic interactive generation of products at the request of users; the development potential is provided related to the virtualization of data sources and processing procedures, the organization of highly efficient distributed processing of large amounts of remote sensing information.

One of the main tasks of the geoportal is to effectively provide users with data on the state of the geophysical fields of the atmosphere at arbitrary points in time. Obviously, such a problem cannot be solved by calculating the final products in advance for all possible query options. Therefore, the geoportal should have tools for the dynamic generation of products based on some basic data set. The basic data set is stored in the geoportal database, and the final products have a limited lifetime in a virtual user environment, from where they can be written to a file on the user's computer.

The basic data set is not identical to the original data and products of satellite radiometry monitoring. In this case, the full processing cycle would take a significant amount of time (for example, interpolation processing of the annual amount of information now requires about a day of calculations for each type of product). In a number of urgent remote sensing tasks (operational monitoring of atmospheric disasters, the study of synoptic processes and climatic changes), it is essential to minimize the time losses associated with the information support of users, either due to the rapid development of the observed processes, or because of the need for stream processing of very large data arrays. Therefore, the basic data set should, on the one hand, be a product of sufficiently deep preliminary processing of the source data, and, on the other hand, with reasonable restrictions on the volume, provide

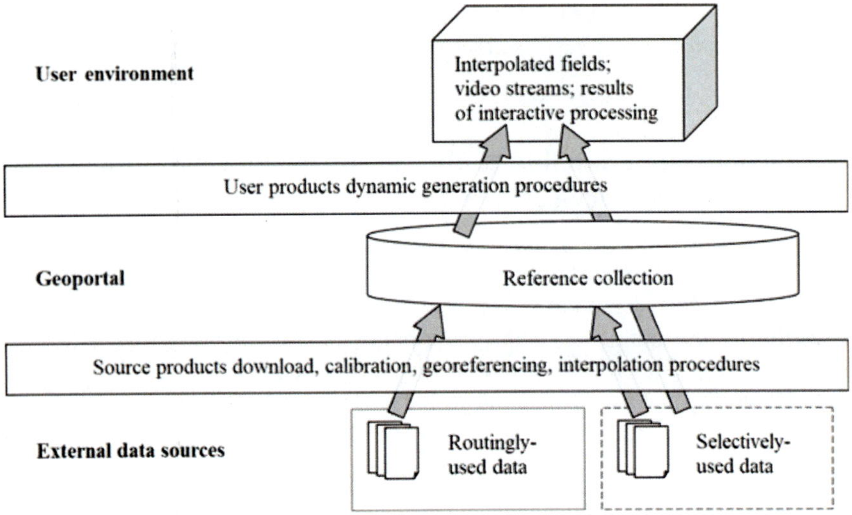

Fig. 7.1 The basic architectural diagram of the geoportal; the transition between data levels corresponding to external sources, geoportal servers and user environment is carried out using specialized geoportal procedures

the possibility of reuse for efficient generation of end-user products. Further, such a basic data set, supplemented with the necessary metadata, is called the "reference collection".

The foregoing is illustrated in Fig. 7.1. It presents three levels of data processing: the level of external sources; geoportal level; user level. The transition between levels is carried out using appropriate software tools (procedures) of the geoportal. It should be noted that these levels do not correspond to generally accepted NASA or CEOS classifications, e.g. (Asrar and Greenstone 1995; Loupian and Savorskiy 2012), but reflect the specifics of the geoportal, as already at the first stage of processing, data of various (usually third or second) levels according to standard classifications can be assimilated.

The level of external sources is represented by freely distributed satellite remote monitoring products. The association of this data with the geoportal is conditional: it is expressed by the fact that the geoportal is equipped with procedures for loading, decoding, spatiotemporal binding (blending) and calibration of this data, which allow full automation. Moreover, part of the data is used on a priority and regular basis to replenish the reference collection using interpolation procedures. Another part of the data can be used selectively: either as additional information in calculating the reference collection, or for the purpose of further joint analysis at the user's request.

The geoportal level corresponds to the contents of its database, including the reference collection. The reference collection is a sequence of global geophysical atmospheric fields on a 0.25° grid, interpolated in 3 h time steps using satellite radiothermovion procedures. Additionally, vector fields of advection observed in the fields of geophysical parameters for the corresponding time intervals are calculated.

At the user level, there are query-generated products. These, in particular, include geophysical fields interpolated at an arbitrary point in time based on data from a reference collection, as well as the results of joint processing of these data with various data from external sources.

Further, the data characteristics of all three levels are considered in more detail.

7.2.1 External Data Sources

The main external data source for the geoportal is currently an open electronic archive of Remote Sensing Systems, USA (Wentz et al. 2012, 2013, 2014a, b). RSS products contain daily fields of reconstructed geophysical parameters of the atmosphere (total precipitable water, cloud liquid water content, etc.) on a 0.25° grid according to satellite scanning radiometers SSM/I, SSMIS, WindSat, AMSR-E, AMSR-2. As the main observations to build the reference collection for different years in the interval 2003–2019 were used data from SSM/I instruments (satellites F13, F14, F15 of the DMSP mission), SSMIS (F16, F17, F18), and WindSat (Coriolis). As additional data, SSM/I (F15), AMSR-E (Aqua) and AMSR2 (GCOM-W1) were used. As an example, Fig. 7.2 shows a diagram of the use of these data for 2004–2016 by Ermakov et al. (2016a). The shaded bands on it show the time ranges of observations in which the data of the corresponding instrument are used on a regular basis. Stripes without shading correspond to available but selectively used data. The vertical lines indicate the narrow time intervals listed in Table 7.1, where such selective addition to the data used on an ongoing basis was required due to significant omissions in the latter.

As can be seen in Fig. 7.2 and in Table 7.1, even over a long observation interval, there are few cases of significant gaps in the regularly used RSS data. In this regard,

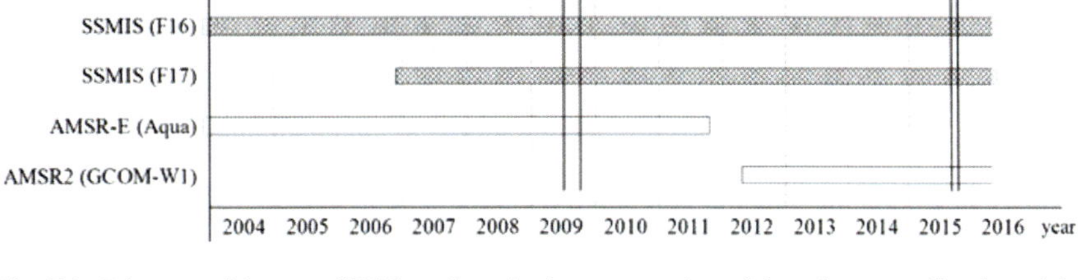

Fig. 7.2 Diagram of the use of RSS products in the construction of the reference collection of the geoportal: products marked with shaded bands are used on a regular basis; vertical numbered lines show the intervals for attracting additional data (white bars), see Table 7.1

Table 7.1 Supplementing main data at problematic intervals shown in Fig. 7.2

Interval		Main data: sensor (platform)	Additional data: sensor (platform)
No	Starting and ending dates		
1	30 Jun 2009–30 Jun 2009	SSM/I (F13), SSMIS (F16, F17), WindSat (Coriolis)	SSM/I (F15)
2	18 Sep 2009–18 Sep 2009	SSM/I (F13), SSMIS (F16, F17), WindSat (Coriolis)	AMSR-E (Aqua), SSM/I (F15)
3	05 Aug 2015–05 Aug 2015	SSMIS (F16, F17), WindSat (Coriolis)	AMSR-2 (GCOM-W1), SSM/I (F15)
4	01 Sep 2015–02 Sep 2015	SSMIS (F16, F17), WindSat (Coriolis)	AMSR-2 (GCOM-W1), SSM/I (F15)

a general approach to the problem of filling such gaps was not developed; each case was considered individually. So, for the data of 2015, an iterative algorithm of satellite radiothermovision was used (Ermakov et al. 2016b). In general, a solution was developed for each case, which made it possible to continuously extend a series of fields of the reference collection in 3 h increments with full coverage of the computational grid over the World Ocean.

A definite drawback of the implemented geoportal tools is the dependence on a single main data source—the RSS archive. It should be noted that this restriction is not fundamental. The functionality of the geoportal can be easily rebuilt to alternative data sources, including domestic satellite devices. Critical indicators in this matter are the quality, volume and regularity of obtaining such data over long observation intervals.

7.2.2 Reference Collection

The methodology for constructing a reference collection from the data of external sources listed above is described in (Ermakov et al. 2015; Ermakov 2018) and the works cited there. The accuracy of the interpolation technique was analyzed (see Chap. 3) and found satisfactory for a wide range of remote sensing problems. When constructing the reference collection, the goal was not to achieve such a high spatiotemporal detailing of the fields, which can be achieved by analyzing relatively short series of observations (Ermakov et al. 2015). The priority was the maximum automation of processing large amounts of data (all cases requiring additional operator intervention are listed in Table 7.1) and a reasonable restriction on the size of the reference collection. On the other hand, a time step of 3–6 h seems sufficient to study the large-scale dynamics of synoptic and climatic processes, and the estimation of all intermediate atmospheric states in three-hour intervals on a 0.25° grid is realized on the basis of data from the reference collection using simple procedures such as optimal interpolation with time (Ermakov et al. 2016b).

According to the type of products, the reference collection is currently subdivided into fields of integral water vapor (total precipitable water, TPW), cloud liquid water content (CLW), and a number of others (Ermakov and Chernushich 2017a), covering a continuous observation interval for 2003–2019. The product range is being expanded. A regular increase in the time coverage of observations across the entire product range is also carried out as the necessary data are accumulated from external sources.

7.2.3 User-Level Products

User-level products are generated dynamically at the request of users based on data from the reference collection and additionally attracted data from external sources. Products are calculated fields on a global grid of 0.25°. The main request parameters are estimated time, product type (TPW, CLW, WND, ADV), type of additional time interpolation (loc, ltw, utc). The last parameter specified controls the dynamic generation of products as follows. When choosing the loc type, the requested field is generated according to the reference collection data that is closest in local time at each grid node to the time specified by the user; additional time interpolation is not performed. When choosing the lwt type for each grid node, the closest local time data is searched from the reference collection before and after the user specified time; then interpolation is performed at a specified point in time. When choosing the utc type, the data of the reference collection are selected and interpolated so that the values at all nodes of the grid correspond to the same universal time (UTC), which coincides with the time specified by the user on the prime meridian. Thus, dynamically generated are both "time scans", which can be interpreted as the results of measurements by some virtual satellite in the solar-synchronous orbit, and the instantaneous global states of the requested geophysical parameter at any time.

7.2.4 User-Level Product File Output Format

The final products are distributed in the form of raster image files containing a calibrated data matrix, supplemented with a standard header in front to directly visualize data in the color palette. The size of the header is 1078 bytes. The size of the data matrix is 720 rows by 1440 columns; the value of B at each node of the grid 0.25° is encoded with one byte and reduced to a physical value in absolute units using linear normalization. Technical details of calibration, spatiotemporal reference, and file format of products are published in (Ermakov et al. 2016a) and the Methodology section on the geoportal website (https://fireras.su/tpw/Radiothermovision.aspx).

7.2.5 *Network Services*

Currently, the geoportal is equipped with basic network services that are typologically relevant to standard tools for previewing, searching and ordering data. Viewing functions correspond to the dynamic visualization of data in the streaming video mode viewed in the video player. The usual functions are provided for starting playback (in 3 h increments per video frame), pausing playback, and setting to an arbitrary point in time. Additional controls allow selecting the year and type of product (TPW, CLW, WND, ADV). It can play in full screen format. Image quality is optimized dynamically depending on the bandwidth of the connection. The visualization mode allows performing preliminary express analysis of the data, the selection of time intervals, zones of interest, the selection and localization of specific objects of study.

The function of searching and ordering fields is supplemented by controls that allow to more accurately set the time of observations, see Fig. 7.3. After selecting the time and type of product, the user initiates the search and data generation by choosing the type of interpolation (loc, ltw, utc, see above) using the buttons to the

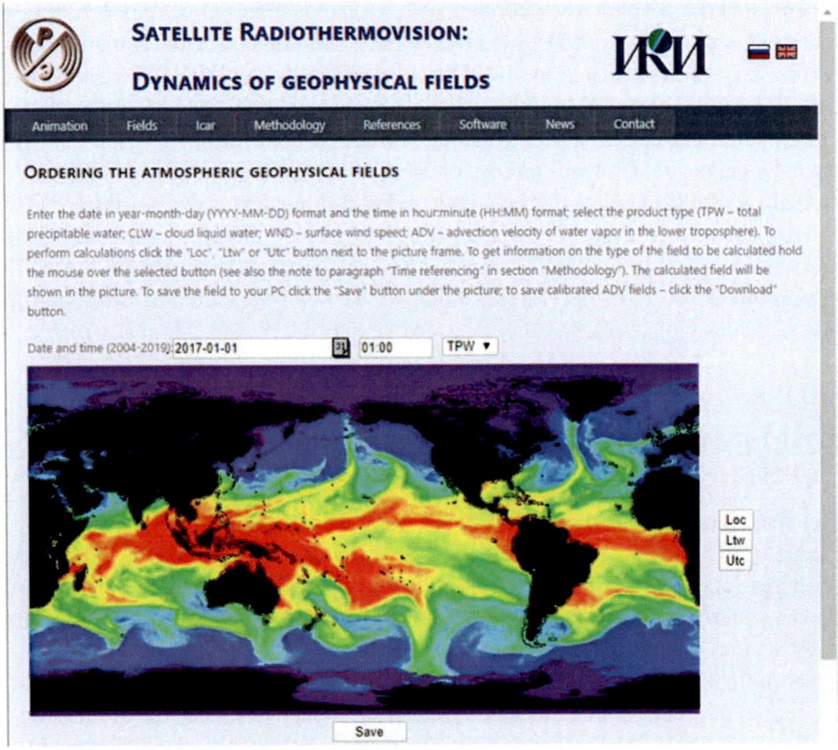

Fig. 7.3 The order page for custom products (interpolated geophysical atmospheric fields over the World Ocean) of the geoportal of satellite radiothermovision

right of the image window. The end product is displayed in the image window and can be saved to a file on the user's computer by pressing the "Save" button.

The architectural orientation of the geoportal to the dynamic generation of user products and the implemented basic tools open up prospects for the development of fundamentally new network services, both in terms of expanding the range of end products and in terms of organizing the work of remote users with distributed remote sensing data. As an example, the next section discusses the implemented and functioning online service of ICAR.

7.3 ICAR Network Service: Remote Processing of Virtually Integrated Data

This section describes the interface and technological principles underlying the ICAR project. An in-depth review of the individual details of the software-algorithmic implementation contains the work (Ermakov and Chernushich 2017b). An essential problem of traditional interface and software solutions used in the organization of geoportals is the provision of functional scalability (Resch and Zimmer 2013). The most widespread approach is that restricts data processing capabilities to some fixed set of tools. Due to this limitation, it is possible, on the one hand, to guarantee correct operation with the initial and expandable data set during the pipeline processing, and on the other hand, to provide each of the tools with a convenient graphical user interface. However, such a restriction seems unnecessarily stringent. It hinders not only the introduction of its own processing procedures into the analysis, but also, in some cases, the adequate adjustment of existing ones.

This problem is especially acute in the presence of complex highly specialized procedures that are of most interest in a deep thematic data analysis. As a rule, the decomposition of such procedures is complex and/or inefficient, since the results of the intermediate processing do not make sense outside the context of a specific task. On the other hand, as the processing algorithm becomes more complex, the number of its potentially required modifications increases almost unlimitedly, which sharply reduces the efficiency of using standard interfaces. A possible approach to overcoming the outlined problem was proposed in the ICAR project, focused on integrated thematic processing of distributed geodata.

7.3.1 Interface Solution

One of the project priorities was the development of a sufficiently flexible interface that allows describing and modifying not only standard tools for analyzing large data arrays (slices, subsets, distribution histograms, statistics, etc.), but also the complex relationships between different types of data presented in the analytical form.

Common graphical user interfaces, as described above, do not provide the necessary degree of flexibility. The universal representation of the algorithm from the analytically described calculation steps is the program code in some programming language. The class of solutions, which considers as an object of network exchange not only data, but also the procedures for their development, corresponds to the most modern trends in the development of distributed information processing technologies, see, for example (Savorskiy et al. 2014), and the literature cited therein. The ICAR project was aimed at developing the simplest to use solution that does not require programming skills from the user. As a result, the idea of a calculator over tabularly described data was proposed and implemented. It was based on the assumption that all spatial data potentially used in the ICAR project is time-bound and can be interpolated without a significant loss of accuracy and information content onto a common computational grid. The grid has 1440 nodes horizontally and 720 nodes vertically and gives a complete coverage of the Earth with a constant pitch between nodes of 0.25°. The calculations specified by the formula entered in text form are performed independently at each grid node. As a result, an array of values is generated, defined on the same computational grid. For ease of visualization, the totals using a linear transformation are reduced to a range from 0 to 250 and rounded to the nearest integer, after which they are displayed on the screen as a raster image with a given color palette and can be saved on a user's computer in a standard format bmp file. Calibration coefficients are stored in the same file, as described in detail in the documentation on the ICAR project page (https://fireras.su/tpw/AboutIcar.aspx), for the possibility of counting back to calibrated values (Ermakov and Chernushich 2017a).

Thus, as in the standard calculator, it is possible to organize calculations by analytically describing them as a text string (a set of strings). The first fundamental extension of the basic functionality is the possibility of introducing into the expression, along with constants, variables of two types—simple scalar variables (for example, containing line numbers, time stamps, etc.) and three-dimensional arrays (for example, fields of geophysical parameters at a given point in time). This possibility is illustrated by two elementary examples.

As a first example, an equation with a simple scalar variable is considered. Her entry in ICAR is as follows:

$$R[x, y, t] = x, \tag{7.1}$$

where R is the name of the result (some allowed combination of letters and numbers), square brackets mean that the previous variable describes the array (the result is defined in all grid nodes for the estimated time), is the column number (horizontal coordinate of the grid cells), is the row number (vertical coordinate of grid cells), is the estimated time (specified by the user using an interface element such as a standard calendar and is calculated in days and fractions of the day from midnight 1 Jan 1990). Expression (7.1) equates the result at each grid node to the horizontal coordinate of the corresponding node. After renormalization, the values monotonically increase

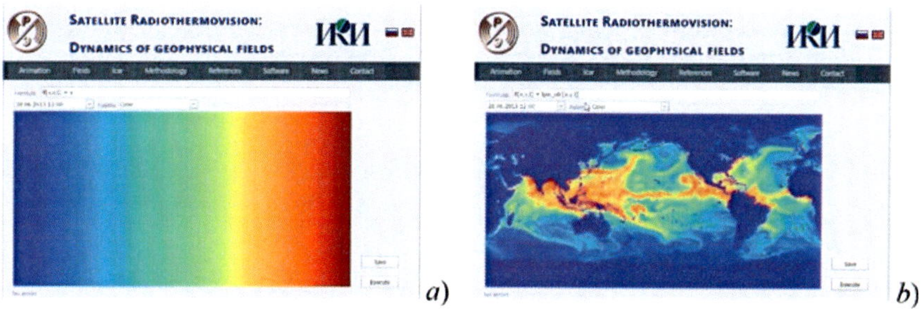

Fig. 7.4 Visualization of the result of the calculation **a** by the Eq. (7.1); **b** by the Eq. (7.2)

horizontally from 0 to 250, so that, in fact, the color scale used will be visualized on the screen (Fig. 7.4a).

As a second example, an equation with a three-dimensional array is considered:

$$R[x, y, t] = \text{tpw\_ utc}[x, y, t]. \tag{7.2}$$

In this case, the R value in each node coincides with the TPW value of the atmosphere calculated in the same node and for a user-specified time instant (Fig. 7.4b). The name of the array (in this case) is unique for each type of product, which, in turn, is determined not only by the type of geophysical parameter, but also by its origin (data source), spatial-temporal binding method, etc. This question is considered in more detail in (Ermakov and Chernushich 2017b).

Other important extensions to the syntax of expressions in ICAR are the conditional operator (branch operator) and the ability to define new functions by the user. The branch operator corresponds to a widely used record with a curly bracket, in which the process of calculating the final result is controlled by one or more conditions. So, for example, the following equation in the syntax of ICAR

$$R[x, y, t] = (x > y)?x : y \tag{7.3}$$

means

$$R[x, y, t] = \begin{cases} x, x > y \\ y, x \leq y \end{cases} \tag{7.4}$$

The ability to introduce user-defined functions is a standard way to expand the syntax of expressions, aimed at simplifying the recording of complex expressions, including those that repeatedly use the same type of calculation sequence. The following equation is considered as an example below:

$$\begin{aligned} &R[x, y, t] = (d(x, y, t) > 0.2)?d(x, y, t) : 0^\circ; \\ &d(x, y, t) = rss\_sst[x, y, t - 100]. \end{aligned} \tag{7.5}$$

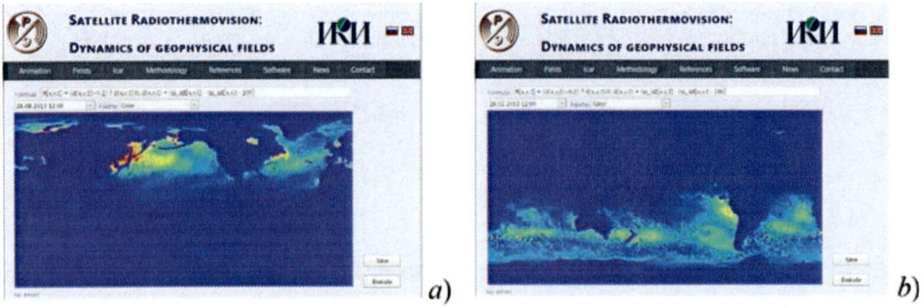

Fig. 7.5 Visualization of the calculation (7.5) for the dates: **a** August 28, 2013; **b** February 28, 2013

The calculation result is equal to the difference between the ocean surface temperatures (rss_sst) on a given date and 100 days before it at the same point if this difference exceeds 0.2 °C, otherwise equals 0. For compactness, a temperature difference function $d(x, y, t)$ is introduced, which first is used in the expression and only then is defined explicitly (after the semicolon). Figure 7.5a illustrates the result of calculations (7.5) for a given current date 08/28/2013. The user has the ability to change the date using a standard interface element such as a calendar, which makes it possible to use the same calculation formula for an arbitrary point in time (if the corresponding satellite or other input information is available). Figure 7.5b illustrates a new calculation result according to (7.5) when changing the date to 02/28/2013.

The next section provides a formal description of the class of text strings interpreted by ICAR.

7.3.2 Syntax and Semantics of ICAR

The description of the syntax and semantics of text strings interpreted in ICAR is conveniently combined on the basis of syntactic graphs (Wirth 1976). A number of formal features are specified in more detail in (Ermakov and Chernushich 2017b).

The text interpreted by ICAR (a sequence of characters) consists of a basic *equation* that relates the result to an *expression* and, possibly, one or more definitions of *functions*, each of which is similar to the main equation, but instead of symbolic designation of the result, begins with a symbolic designation of the function previously used in the expressions. Each new function definition is separated from the preceding text by a semicolon. Thus, the entire text consists of a set of lines separated by a semicolon, with the first line being an *equation*, and the rest (if any) are *function* declarations. Therefore, it is sufficient to provide complete descriptions of the syntax and semantics of the *equation* and *function* separately.

The diagram in Fig. 7.6 gives a complete description of the line, corresponding to the concept of the *equation* in ICAR. According to the method described by Wirth,

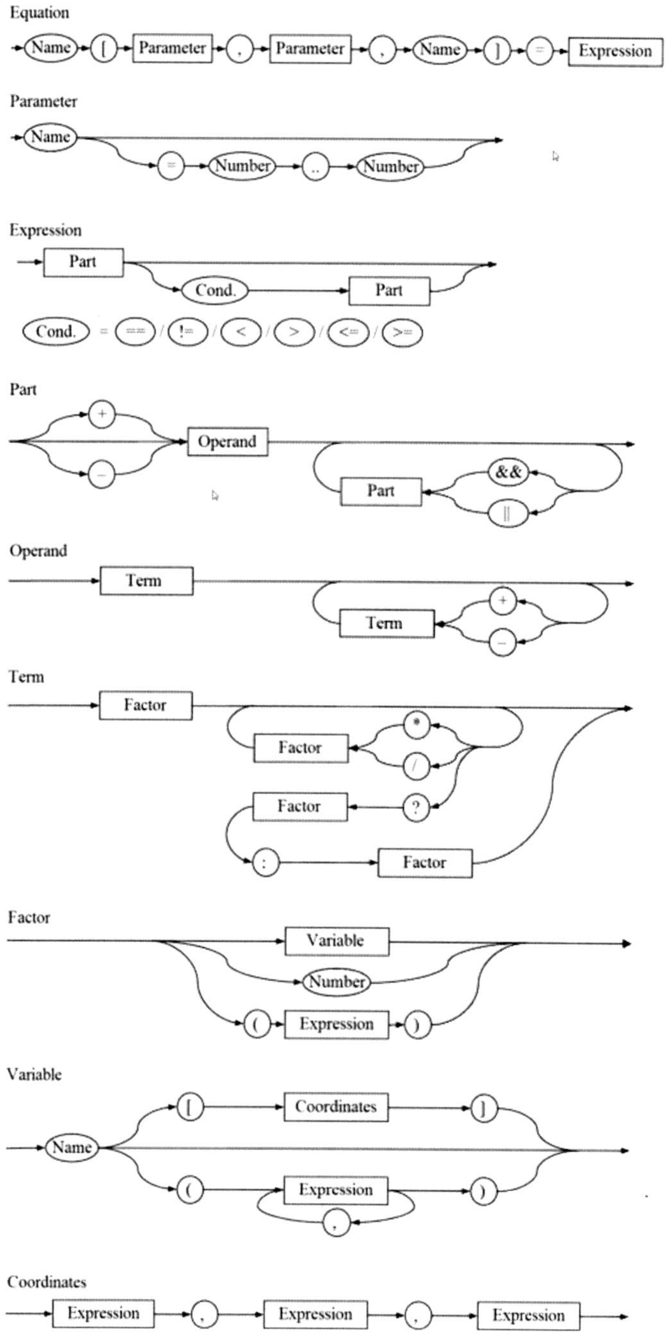

Fig. 7.6 Syntactic graph for the concept of *equation* in ICAR

the refinement goes in the direction from general to particular, while elementary terms (tokens) are presented in oval frames, and compound ones in rectangular ones.

As shown in Fig. 7.6, the *equation* sequentially lists: the symbolic name of the result, its *parameters* in square brackets, and the actual calculated *expression* after the equal sign. The parameters contain the names of scalar variables corresponding to the column number (node number in the row), row number, and time. For the first two variables, it is also possible to specify a range of changes (minimum and maximum values). By default, the range of changes covers the entire calculation grid. Variable names are assigned by the user arbitrarily and can be further used in the calculated expressions.

In turn, the *expression* consists of one *part*, or of two *parts*, united by a logical *condition*: "equal", "not equal", "less", "more", "less than or equal", "more or equal". In the latter case, the *expression* is considered equal to one if the condition written in it is satisfied, and zero if the condition is not fulfilled.

Part of the expression is defined recursively. It can consist of one *operand* (possibly with the plus or minus sign preceding it) or combine with other *parts* using logical conjunction ("AND", "&&") and/or disjunction ("OR", "||") operators. The parsing of the components is carried out until they are completely decomposed into operands, combined by logical operations.

An *operand* is a *term* or several *terms* connected by addition and/or subtraction operators.

The *term* is a *factor* or several *factors* combined by the operators of multiplication and/or division, and can be additionally completed by a branch operation consisting of the symbol "?", the *factor*, the symbol ":" and the alternative *factor*. The rule for calculating the result in the branch operation is as follows: if the value of the *factor* preceding the "?" sign is different from zero, then the result is the value of the *factor* immediately following the "?" sign; otherwise, the result is the value of the *factor* following the sign ":" (alternative), see also the description for Eqs. (7.3) and (7.5).

The *factor* may be a *variable, number* or a new *expression* in parentheses.

A *variable* can be represented by one *name* (corresponds to a simple scalar value), or a *name with coordinates* in square brackets (corresponds to a three-dimensional array), or a *name* with an *expression* or several *expressions* separated by commas in parentheses (corresponds to the function of one or more variables).

Coordinates are a sequence of three *expressions* separated by commas.

Thus, parsing an *equation* according to the rules shown in the diagram in Fig. 7.6 allows to either represent the final result as a composition of logical and arithmetic operations on the values of variables, arrays, and functions, or to reveal a syntax error in it. The priority of operations is also mainly determined by the indicated syntax rules: the highest is for operations of division and multiplication, lower is for operations of addition and subtraction, even lower is for logical operations of conjunction and disjunction (it is additionally accepted that the priority of the first one is higher), the most low—for comparison operations. Equal priority operations are performed in the order listed in the expression (from left to right).

To close the computational algorithm, it is necessary to additionally determine the method of obtaining input data values—simple variables, arrays, and functions.

It is accepted that in ICAR expressions only simple variables can appear that correspond to the result parameters (node number, line number and time point) or formal parameters of the function (inside its definition). Moreover, during the calculations, the values of the node number and line number sequentially run through the ranges of allowed values, and for each combination of node and line numbers, the calculation is performed according to the equation. The moment of time is fixed before the start of calculations and can be set by the user using the ICAR graphical interface, as described above. The initialization of the function parameters occurs before its call from the calculated expression. Thus, at any moment of calculations, the values of all simple variables are uniquely determined.

Arrays in ICAR are families of matrices parameterized by a moment in time. For a user-specified point in time, each array is a set of values defined in all cells of the computational grid 1440 (columns) × 720 (rows). When setting a new point in time, the array values, generally speaking, require updating. An array is a convenient way to describe a two-dimensional field (the field of a geophysical parameter) that varies over time. A unique array name is used to indicate a specific geophysical parameter (Ermakov and Chernushich 2017b). Thus, the use of arrays with names from a limited, predetermined set, each of which corresponds to a specific product, is permitted in the ICAR equations. The product can either be stored on the server of the geoportal of satellite radiothermovision, or be the result of dynamic generation and/or virtual integration of data from other archives, satellite data bases or geoportals. To ensure uniform inclusion in the equations of any types of products, a unified tool for loading products in the process of their use in the calculations is implemented.

ICAR allows the use of two types of functions: library and user. Library functions have predefined names and a fixed (generally speaking, different for different functions) number of parameters. The full set is described on the project website (https://fireras.su/tpw/AboutIcar.aspx).

Custom functions are one of the traditional ways to extend the syntax for a more compact and convenient description of the calculation procedure, as well as for organizing non-trivial methods of calculation, for example, recursion. The peculiarity of the ICAR syntax is that it is allowed to use functions in calculated expressions before its formal definition. An absolute requirement is the strict correspondence of the function call to its definition (i.e., the exact coincidence of the function name and the number of parameters). For an example, see the description of formula ((7.5). The definition of the syntax and semantics of a function in ICAR is relatively simple and is described in (Ermakov and Chernushich 2017b).

7.3.3 *Software Implementation*

The software implementation of ICAR is based on the principle of partial compilation and computations on the stack, as described in detail in (Ermakov and Chernushich 2017b). Partial compilation consists in translating the main formula and function definitions into a stack representation and creating name tables. Execution is carried out using software interpreters operating on the principle of stacked machines. At the beginning of execution, two nested loops are organized. The outer loop changes the current value of the line number (second argument of the result) from the minimum value to the maximum. The inner loop changes the column number values. For each combination of column and row numbers, a program interpreter related to the main equation is called. The interpreter's temporary memory, intended for storing parameters (row and column numbers and time points), their current values are entered. The software interpreter makes copies of its stacks of operations and data (since they will change during execution and must be restored when called with the following set of parameters). Then, the program interpreter removes the first operation from the top of the operations stack and begins the recursive calculation of the result according to the scheme described above. The calculated value is entered into the array element in temporary memory, reserved for storing the result. The item index is determined by the current column and row numbers.

When the interpreter reaches the invocation of the user-defined function, execution continues recursively. First, the number of arguments is determined and their values are calculated. Next, control is transferred to a new program interpreter working with copies of the stacks of operations and data (and a table of names) built for the corresponding function. In this case, the temporary memory of the interpreter is initialized with the calculated values of the function parameters. Upon receipt of the final result, control is returned to the program interpreter that initiated the function call. The library function interpreter does not work with stacks. The procedure for calculating the result in this case is rigidly coded in the program, and the interpreter calls it with the parameter values passed to it.

When the interpreter reaches the operation of obtaining the value of the array, the corresponding data loader is called. At the beginning, the loader checks to see if the requested data has already been downloaded previously. At the first request, the loader calls the download procedure specific to each data type. In general, downloading can consist of establishing a connection with a remote server using one of the standard network exchange protocols (ftp, http, etc.), obtaining the required data files, unpacking them, decoding them, and spatiotemporal binding with interpolation to the general design the grid. Thus, the data request procedure can be a significant part of the execution time, and it is performed multiple times (at least once for each possible combination of grid node indices). Therefore, all the data downloaded in a single work session is temporarily stored on the geoportal server and immediately ready to work with a second request. Checking the readiness of the requested data consists in constructing a unique data identifier, by the name of the array (data type) and the current time, and in the search for the address in the temporary memory associated with this identifier. If such an address is found, the loader immediately accesses the data at this address. Otherwise, the loader performs the complete downloading procedure, builds a new data identifier and associates it with the memory allocated for newly loaded data.

7.3.4 Data Provisioning and Functional Content

As indicated above, at present, various types of data arrays can be used in ICAR expressions. Their most relevant description is contained on the ICAR project website (https://fireras.su/tpw/AboutIcar.aspx).

The main source of data for calculations in ICAR is currently the geoportal of satellite radiothermovision. However, thanks to the unified architecture of the loaders, it is possible to use data from other archives, databases and geopotals according to the principle of virtual integration (Ermakov and Chernushich 2017a).

Of course, an important aspect of the project is the provision of a wide range of library analytical functions. The most relevant list of them is also contained on the ICAR project website (https://fireras.su/tpw/AboutIcar.aspx).

The possibility of introducing new library functions is provided by the architecture of the software interpreter. Adding a function is carried out by integrating a new specific software interpreter into the system. This ensures an unlimited expansion of the functionality of ICAR and the geoportal of satellite radiothermovision.

7.4 The Practice of Using Implemented Software Solutions

The universality of the approach laid down in the architecture of the geoportal of satellite radiothermovision and the ICAR project is manifested in the possibility of organizing calculations that are characteristic of most geoportals (various statistical characteristics, profiles, data slices, time series, etc.), as well as specific, aimed at solving specific practical tasks. It is important that in the latter case the necessary flexibility is maintained, which makes it easy to make arbitrary modifications to the calculated expressions in an interactive mode, almost immediately tracking and comparing the results of the changes made. Thus, the ICAR project can be used both for primary data analysis, and for the development, modification and debugging of an algorithm for their processing, as well as for organizing stream processing of large data arrays using a developed algorithm.

For effective streaming processing, a network service (web-service) has been implemented on the geoportal of satellite radiothermovision, which opens a functional interface to ICAR. The function sends the calculation equation and estimated time to the server, and receives the result in the form of an array on the calculation grid or a text message about the error. Thus, it is possible to develop a client application that performs streaming processing using the functionality of the ICAR software kernel (Ermakov and Chernushich 2017a).

It is important to note the peculiarity of network data exchange during the implementation of calculations in ICAR, consisting in the fact that network traffic between the user's computer and the geoportal server is minimized. The user transmits to the server only the text of the main equation and function definitions and receives only an array of calculated values or a text error message. Virtual data integration is carried out through network exchange between the geoportal server and distributed archives and does not affect the amount of traffic transmitted to the user. It is also important that the text transmitted by the user is at the same time a way of documenting (describing and saving) the processing algorithm. This text can be written to a text file and subsequently reused both in an interactive session and with the help of a client program.

The work (Ermakov and Chernushich 2017b) contains an illustration of the practical use of ICAR by the example of calculating the so-called "Temperature-Humidity criterion" (Rostovtseva and Goncharenko 2014). Other examples of practical applications are discussed in the context of atmospheric river research (Ermakov 2017).

The implementation of streaming processing using the functional tools of the geoportal enables a large-scale statistical analysis of the various proposed characteristics and criteria over long, climatically significant intervals of satellite radiometry observations.

7.5 Concluding Remarks to this Chapter

1. The research results presented in this Chapter relate to the development and creation of a geoportal of satellite radiothermovision.
2. The developed and created geoportal of satellite radiothermovision (https://fireras.su/tpw/) implements remote work procedures with global fields of a number of geophysical characteristics of the ocean-atmosphere system (total precipitable water, cloud liquid water content, near-surface wind speed, etc.) in the continuous observation interval of years 2003–2019 with full spatial coverage over the World Oceans on a 0.25° grid, as well as interactive joint processing of these data with information from other open sources by means of virtual integration.
3. Prospects for the use of the geoportal are associated, in particular, with the possibility of effective mutual calibration of data from various satellite devices, as well as integrated express analysis of atmospheric processes using the implemented procedures for dynamic processing and visualization of data integrated into the portal. The geoportal can be used as an informational source of the data on detailed atmospheric dynamics in studies of various mesoscale and synoptic processes, and global atmospheric circulation as a whole.

References

Asrar G, Greenstone R (eds) (1995) MTPE/EOS reference handbook. NASA pub. NP-215. National Aeronautics and Space Administration, Washington, DC 276 p

Ermakov DM (2017) Investigation of the features of long-term global atmospheric circulation via satellite radiothermovision. In: Proceedings of 2017 progress in electromagnetics research symposium—Spring (PIERS), pp 413–418. https://doi.org/10.1109/piers.2017.8261775

Ermakov DM (2018) Global circulation of latent heat in the Earth's atmosphere according to data from satellite radiothermovision. Izv Atmos Ocean Phys 54(9):1223–1243

Ermakov DM, Chernushich AP (2017a) Current capabilities of the geoportal of satellite radothermovision and some results of the ICAR project. Sovremennye Problemy Distantsionnogo Zondirovaniya Zemli iz Kosmosa (Curr Probl Remote Sens Earth Space) 14(7):46–57. https://doi.org/10.21046/2070-7401-2016-13-3-46-57 (in Russian)

Ermakov DM, Chernushich AP (2017b) Development of network services of the geoportal of satellite radio thermal vision. Elektronnyye biblioteki (Electron Libr) 20(1):50–76. https://elbib.ru/article/view/401 (in Russian)

Ermakov DM, Raev MD, Suslov AI, Sharkov EA (2007) Electronic long-standing database for the global radiothermal field of the Earth in context of multy-scale investigation of the atmosphere-ocean system. Issledovanie Zemli iz kosmosa (Earth Res Space) (1):7–13 (in Russian)

Ermakov DM, Sharkov EA, Chernushich AP (2015) Satellite radiothermovision of atmospheric mesoscale processes: case study of tropical cyclones. Int Arch Photogramm Remote Sens Spat Inf Sci ISPRS Arch XL(7/W3):179–186

Ermakov DM, Chernushich AP, Sharkov EA (2016a) Geoportal of satellite radiothermovision: Data, services, prospects. Sovremennye Problemy Distantsionnogo Zondirovaniya Zemli iz Kosmosa (Curr Probl Remote Sens Earth Space) 13(3):46–57. https://doi.org/10.21046/2070-7401-2016-13-3-46-57 (in Russian)

Ermakov DM, Sharkov EA, Chernushich AP (2016b) A multisensory algorithm of satellite radiothermovision. Izv Atmos Ocean Phys 52(9):1172–1180

Ermakov DM, Sharkov EA, Chernushich AP (2017) Satellite radiothermovision on synoptic and climatically significant scales. Izv Atmos Ocean Phys 53(9):973–978

Gorelick N, Hancher M, Dixon M, Ilyushchenko S, Thau D, Moore R (2017) Google earth engine: planetary scale geospatial analysis for everyone. Remote Sens Environ 202:18–27

Lee CA, Gasster SD, Plaza A, Chang CI, Huang B (2011) Recent developments in high performance computing for remote sensing: a review. IEEE J Sel Top Appl Earth Obs Remote Sens 4(3):508–527

Loupian EA, Burtsev MA, Proshin AA, Kobets DA (2018) Evolution of remote monitoring information systems development concepts. Sovremennye Problemy Distantsionnogo Zondirovaniya Zemli iz Kosmosa (Curr Probl Remote Sens Earth Space) 15(3):53–66. https://doi.org/10.21046/2070-7401-2018-15-3-53-66 (in Russian)

Loupian EA, Savorskiy VP (2012) Basic products of Earth remote sensing data processing. Sovremennye Problemy Distantsionnogo Zondirovaniya Zemli iz Kosmosa (Curr Probl Remote Sens Earth Space) 9(2):87–96. (in Russian)

MIMIC-TPW (2012) Morphed integrated microwave imagery at CIMSS—total precipitable water (MIMIC-TPW) (electronic resource). URL: http://tropic.ssec.wisc.edu/real-time/mimic-tpw/global2/main.html

MIMIC-TPW2 (2020) Morphed integrated microwave imagery at CIMSS—total precipitable water. Version 2. (MIMIC-TPW ver.2). URL: http://tropic.ssec.wisc.edu/real-time/mtpw2/product.php?color_type=tpw_nrl_colors&prod=global2&timespan=24hrs&anim=html5

Resch B, Zimmer B (2013) User experience design in professional map-based geo-portals. ISPRS Int J Geo Inf (2):1015–1037

Rostovtseva VV, Goncharenko IV (2014) Evaluation of the influence of thermal stratification of the troposphere on the activity of tropical cyclogenesis according to satellite microwave radiometry. Issledovaniye Zemli iz Kosmosa (Earth Res Space) (2):3–17. (in Russian)

Savorskiy V, Lupyan E, Balashov I, Proshin A, Tolpin V, Ermakov D, Chernushich A, Panova O, Kuznetsov O, Vasilyev V (2014) Basic technologies of web services framework for research, discovery, and processing the disparate massive Earth observation data from heterogeneous sources. ISPRS Arch 40(4):223–228

Wentz FJ, Hilburn KA, Smith DK (2012) Remote sensing systems DMSP SSM/I, SSMIS daily environmental suite on 0.25° grid, version 7, 8 (electronic resource). Remote Sensing Systems, Santa Rosa, CA. URL: http://www.remss.com/missions/ssmi/

Wentz FJ, Ricciardulli L, Gentemann C, Meissner T, Hilburn KA, Scott J (2013) Remote sensing systems coriolis WindSat daily environmental suite on 0.25° grid, version 7.0.1 (electronic resource). Remote Sensing Systems, Santa Rosa, CA. URL: http://www.remss.com/missions/windsat

Wentz FJ, Meissner T, Gentemann C, Brewer M (2014a) Remote sensing systems AQUA AMSR-E daily environmental suite on 0.25° grid, version 7 (electronic resource). Remote Sensing Systems, Santa Rosa, CA. URL: http://www.remss.com/missions/amsr

Wentz FJ, Meissner T, Gentemann C, Hilburn KA, Scott J (2014b) Remote sensing systems GCOM-W1 AMSR2 daily environmental suite on 0.25° grid, version 7.2 (electronic resource). Remote Sensing Systems, Santa Rosa, CA. URL www.remss.com/missions/amsr

Wimmers AJ, Velden CS (2011) Seamless advective blending of total precipitable water retrievals from polar orbiting satellites. J Appl Meteorol Climatol 50(5):1024–1036

Wirth N (1976) Algorithms+data structures = programs. Prentice-Hall, Englewood Cliffs, NJ, 400p

Yao X, Li G, Xia J, Ben J, Cao Q, Zhao L, Ma Y, Zhang L, Zhu D (2020) Enabling the big earth observation data via cloud computing and DGGS: opportunities and challenges. Remote Sens 12(1):62 (1–15). https://doi.org/10.3390/rs12010062

Chapter 8
Conclusion

The book describes a new approach to the processing and interpretation of the data of satellite radiothermal monitoring of the Earth, called satellite radiothermovision. Unlike the traditional paradigm, which treats these data as independent point instantaneous measurements, the approach is aimed at extracting additional information contained in them, due to the presence of spatiotemporal relationships on the scales of the corresponding atmospheric processes (Chap. 2).

Due to the implementation of this approach, it is possible to obtain a more complete picture of the state of the atmosphere and its dynamics. First of all, it becomes possible to partially fill in the gaps in the data relating to the regions and the points in time not covered by satellite observations. Further, it becomes possible to reconstruct the dynamics of the atmosphere, retrieving the speed and direction of air advection.

As a result, it is possible to calculate a number of geophysical parameters that were not previously retrieved directly from remote data, the most important of which is the power of the latent heat flux integrated over the height of the atmosphere. This characteristic is significant, and sometimes determining, in the formation and development of many atmospheric mesoscale and synoptic processes (Chap. 3).

Because of natural limitations, only some of the indicated types of processes were selected as objects of study when demonstrating the practical application of the satellite radiothermovision method.

Thus, a clear connection has been shown between the convergence and divergence of atmospheric latent heat fluxes with the intensification and dissipation of tropical cyclones. At the same time, the characteristic values of the maximum flux power (units of petawatts) are in good agreement with theoretical estimates, many times exceed the total vertical heat flux from the ocean to the tropical cyclone and in general explain its overall energy balance (Chap. 4).

New possibilities of the approach are demonstrated in the study of large-scale synoptic processes, in which one of the methodological difficulties was the need to simultaneously observe extended regions of the Earth without data omissions. The

D. M. Ermakov, *Satellite Radiothermovision of Atmospheric Processes*, Springer Praxis Books,
https://doi.org/10.1007/978-3-030-57085-9_8

advantages of the new approach are illustrated by an example of the study of atmospheric rivers—specific phenomena that not only sporadically form extreme weather conditions on the coasts and inland, but appear to be an important and intrinsic element of the general circulation of atmospheric moisture and heat (Chap. 5).

The retrospective application of the satellite radiothermovision approach to the analysis of long-term archival data of remote observations allows studying the most large-scale properties of the atmosphere as the Earth's climate subsystem. This is shown by the example of reconstruction of the structure and parameters of global atmospheric circulation. It should be noted that, based solely on observational data, the method of satellite radiothermovision is a method of obtaining information on climate trends, independent of numerical climate models and ground-based observations. This has the rich potential of new opportunities for intercomparison with model scenarios, for verification and refinement of up-to-date ideas about climate trends (Chap. 6).

As noted above, it is impossible to cover all the variety of types of atmospheric processes in the analysis. It is extremely important to ensure a wide scientific community with free access to remote sensing data processing products based on the method of satellite radiothermovision. This may turn out to be useful both for checking the conclusions made earlier in new case studies, and, primarily, for an in-depth study of various atmospheric processes, including those not considered in the book. To this end, the author and his colleagues have developed and maintained a geoportal of satellite radiothermovision, which is described in Chap. 7 of the book.

Work on improving the proposed method continues uninterruptedly. A number of new results, for various reasons, remained outside the scope of this book. In conclusion, let us briefly outline some ways of further development that seem to be the most significant.

8.1 Improving the Quality of Dynamic Analysis Products

As mentioned in Chap. 3, it is simple enough to verify that the interpolated fields of the geophysical parameters of the atmosphere are in good agreement with those reconstructed from independent satellite observation data. But this only indirectly indicates the correct computation of atmospheric dynamics. It is much more difficult to give a numerical estimate of the accuracy of calculating the velocity of advection, since this is an integral over the entire height of the atmosphere. There is no alternative satellite data—neither air velocity nor vertical humidity profiles—that would provide the ability to compare with the required detail and accuracy. Under these conditions, the only way out is to compare with relatively sparse data from weather stations and with reanalysis data. Such work is ongoing.

An important algorithmic step should be the introduction of the criterion of quality of calculation, by analogy with that used in reconstructing atmospheric motion vectors from geostationary weather satellites. The methodological difficulty here is that the pyramidal algorithm used in satellite radiothermovision is an order of magnitude more complicated. Nevertheless, the indication of certain obviously unreliable decisions, as the first version of the quality assessment, can and should be implemented at the next stages of the development of the approach.

8.2 Enhanced Spatial Coverage and Improved Detail

The opportunity of observing the atmosphere only over the ocean sometimes significantly limits the applicability of the method in practice. It is known that many atmospheric processes that form and evolve over the ocean are able to continue their life cycle over land for a fairly long time. In exceptional cases, this applies even to tropical cyclones. This is even more true for atmospheric rivers. The retrieved characteristics of the general atmospheric circulation over land will undoubtedly significantly supplement and refine the picture presented in Chap. 6. Hence, it is relevant to involve in the analysis some new data containing information on the state of the atmosphere over all types of the Earth's surface. Here, undoubtedly, the development of satellite radiometric complexes of a new generation plays a decisive role.

The same can be said about improving spatial resolution of data. As shown in the book, the method of satellite radiothermovision did not reach the limit on the detail of the interpolated fields of atmospheric parameters. Limitations are inherent in the source satellite data and/or the methodology for their preprocessing (for example, reduction to a grid of 0.25 geographic degrees). In cases where higher resolution information is available, the calculation can be successfully performed on grids of 0.2 and 0.125 geographic degrees, and this, apparently, is far from the limit.

Here, it is relevant to note that a certain advantage could be given by the rejection of preliminary interpolation of data on a regular geographic grid with inevitable distortions in this case. The possibility of realizing dynamic analysis directly on the grid of initial measurements in a spherical coordinate system and taking into account the multiple passes of the satellite over high latitude regions within one day is extremely attractive.

8.3 Extension of Analysis Time Interval

With respect to the time interval for satellite data analysis, two diametrically opposite aspects should be noted. One of them is associated with the possibility of obtaining the promptest information on current atmospheric dynamics. In addition to some technical difficulties, the problem here is that in the implemented version of the method, atmospheric dynamics is restored in the interval *between* the fields of two daily observations. Thus, it is necessary to have complete daily observational data *before* restoring the atmospheric dynamics preceding the end of this day. In principle, this limitation can be overcome by implementing extrapolation of the current dynamic picture one day in advance. In practice, at present, the "lag" of the final products of the dynamic analysis from the current date can be up to one year due to the labor required for the periodic updating of information. So, e.g. at the time of writing this book (May 2020), the geoportal of satellite radiothermovision contains interpolated geophysical fields of the atmosphere only until 2019 inclusive. Under these conditions, it is very premature to talk about the real feasibility and quality of extrapolation to one day forward from the current date. However, such a potential feature of the method must be kept in mind.

The second aspect is the extension of the retrospective analysis to earlier dates. The initial year (2003) of the information presented on the geoportal is not accidental. It has been established that the information redundancy of satellite radiometers that operated before 2003 is not enough to reconstruct a sufficiently complete dynamic picture of the atmosphere. However, an important reservation is required here. Only radiometers operating in solar-synchronous orbits are meant. Some progress may be due to the complication of the algorithm for creating reference fields and for motion estimation/compensation, which would make it possible to use for analysis the data from satellites in other orbits (including those with low inclination, since the most significant data gaps occur in tropical belt).

8.4 Extension of the Dimensionality of the Task

Although the book demonstrates the adequacy of the analysis of atmospheric dynamics in a two-dimensional formulation, one cannot overlook the fact that the atmosphere is three-dimensional and most atmospheric processes have a pronounced heterogeneity in height. In this sense, an important stage in development would be the construction of an algorithm for three-dimensional analysis of atmospheric dynamics. It can be noted that this is a significant step from a methodological point of view, since the vast majority of the algorithms used in practice are two-dimensional and are associated with the analysis of "flat" images. Thus, although the practical application of such three-dimensional analysis is currently constrained by the lack of sufficiently complete and accurate observations of the three-dimensional structure of the atmosphere, the problem has obvious prospects and is of undoubted interest. One can begin to solve it, for example, on the basis of a model object of study in the form of a series of reanalysis data that are resampled to the frequency of one or two fields per day.

The above list of tasks is obviously incomplete and reflects the author's opinion on the development strategy of a new approach proposed by him and his colleagues to the interpretation and processing of satellite radiothermal observation data. Even in its current form, the approach has clear prospects for wider practical application, for example, in the study of extratropical cyclones, polar cyclones, regional climate features (including the Arctic and Antarctic), characteristics of the zonal atmospheric circulation, and so on. Concluding this chapter and the book, the author wants to repeat once again that he will consider his goal to be fully achieved if the book gives impetus to the development of any kinds of studies in which the ideas and results described here appear useful.